SECOND EDITION

WEB WISDOM

How to Evaluate and Create Information Quality on the Web

MARSHA ANN TATE

CRC Press
Taylor & Francis Group
Boca Raton London New York

CRC Press is an imprint of the
Taylor & Francis Group, an **informa** business

CRC Press
Taylor & Francis Group
6000 Broken Sound Parkway NW, Suite 300
Boca Raton, FL 33487-2742

Library of Congress Cataloging-in-Publication Data

Tate, Marsha Ann.
  Web wisdom : how to evaluate and create information quality on the Web / author, Marsha Ann Tate.
      p. cm.
  Includes bibliographical references and index.
  ISBN 978-1-4200-7320-1 (alk. paper)
  1. Web sites. 2. Web site development. 3. World Wide Web. I. Title.

TK5105.888.A376 2010
004.67'8--dc22   2009020890

Visit the Taylor & Francis Web site at
http://www.taylorandfrancis.com

and the CRC Press Web site at
http://www.crcpress.com

Dedication

To my mother, Barbara, and in memory of my father, Andrew Tate Jr., and my grandfather, Andrew Tate Sr. Their enduring love and confidence in me made this all possible.

Contents

List of Illustrations

Preface

The World Wide Web has undergone tremendous growth since the first edition of *Web Wisdom: How to Evaluate and Create Information Quality on the Web* was conceived and written in the mid-to-late 1990s. In 1995, there were only 45 million Internet users worldwide; one decade later, the number of Internet users across the globe surpassed the one billion mark and by 2011 the global Internet community is projected to reach two billion users. A number of forces have helped fuel the global Internet revolution, including (1) the development of portable, mobile-based technologies such as smart phones that incorporate Web searching, texting, e-mail, and related capabilities; (2) faster Internet connection speeds; and (3) increased access to computer-based technologies overall. Moreover, the ability for individuals to be "connected" to the Internet 24/7 has fostered a new phenomena, *social media*, an umbrella term that encompasses activities such as blogging, twittering, podcasting, and more. A decade ago, these activities were the purview of small select groups of Internet users or simply did not exist (worldwide Internet users 2006).

Despite the dramatic changes in the online realm over the past decade, the basic evaluation principles presented in the first edition of *Web Wisdom* remain equally applicable today as they did in the late 1990s. Focusing on the authority, accuracy, objectivity, currency, and coverage of content irrespective of format remains a reliable method to assess the quality of information. Unfortunately, as online technologies mature and the use of Internet-based content becomes ubiquitous, many people mistakenly assume there is less need to emphasize critical evaluation skills. On the contrary, the phenomenal global growth of the Internet coupled with the ever-increasing sophistication of online technologies and software applications require individuals to be even more savvy Web users than in the past.

With this in mind, the goal of the second edition of *Web Wisdom* is to demonstrate how to adapt and apply the five core traditional evaluation criteria (authority, accuracy, objectivity, currency, coverage) originally introduced in the first edition to the modern-day Web environment.

On a related note, the book introduces a series of checklists comprised of basic questions to ask when evaluating or creating a particular type of Web page. These checklists can be utilized two different ways based on the reader's preference. First, they can be used similarly to any other checklist, with each question answered in sequential order. On the other hand, the checklists can be used more figuratively, with the questions and their underlying concepts serving as guiding principles rather than as a rigid set of rules.

Acknowledgments

I would like to thank my mother, Barbara Tate, and my friend and colleague, Barbara Coopey, assistant head, Access Services, The Pennsylvania State University Libraries, for their assistance and encouragement throughout the process of writing both editions of *Web Wisdom*. I would also like to thank the following businesses and organizations who have generously granted me permission to use screen captures of their Web pages in the book:

- The Math Forum at Drexel University
- Penn State Public Broadcasting
- The Pennsylvania State University
- Roots Canada Ltd.

About the Author

Marsha Ann Tate received a B.A. degree in political science from The Pennsylvania State University, an M.S. degree in library science from Clarion University of Pennsylvania, an M.A. degree in communication studies from Bloomsburg University of Pennsylvania, and a Ph.D. degree in mass communications from The Pennsylvania State University. Dr. Tate currently works as a librarian and Web site coordinator at the University Park Campus of The Pennsylvania State University. She is also a freelance writer, researcher, and community education instructor. In addition to *Web Wisdom*, Marsha is the author of *Canadian Television Programming Made for the United States Market: A History with Production and Broadcast Data* (McFarland, 2007).

Related Web Site

A companion Web site to *Web Wisdom: How to Evaluate and Create Information Quality on the Web,* 2nd edition, is available at http://mtateresearch.com/web_wisdom/. The resources available at the site include the following:

1. Links to many of the Web page examples used throughout the book as well as links to numerous other sites that illustrate Web evaluation concepts
2. PowerPoint presentations that address topics such as the five traditional evaluation criteria and their application to Web resources, advertising and sponsorship on the Web, and evaluation strategies for social media content
3. A webliography of Web evaluation and other related resources
4. A glossary of Web-related terms
5. Contact information for the author

1 Web Wisdom
Introduction and Overview

INTRODUCTION

The World Wide Web offers us unprecedented communicative powers. It enables us to read breaking stories from news sources around the world, track population estimates on a second-by-second basis, and locate medical information on nearly every disease imaginable. In fact, the Web makes possible the instant retrieval of information on virtually any topic we care to explore.

It is also revolutionizing our buying habits. We can make online plane and hotel reservations and browse through countless virtual stores, purchasing merchandise from our desktops and personal data assistants. Moreover, blogs, wikis, and myriad other Internet and mobile-based networking tools are transforming our social lives. As a whole, our unprecedented access to information and ability to communicate with others on a global scale has fundamentally changed our society. But how, among this extraordinary abundance of resources, do we know what to believe? How can we determine what information is authoritative, reliable, and therefore trustworthy? Although the challenge of evaluating resources is as old as information itself, the Web brings new and sometimes complicated twists to the process. This book provides tools and techniques to help meet the sometimes straightforward and sometimes convoluted evaluation challenges posed by the Web.

Nonetheless, the book is not just directed toward Web users. It also provides important guidance for creators of Web-based resources who have information that they want to be recognized as reliable, accurate, and trustworthy. For example, how can a Web user know whether to trust information from a page or site if the creator does not include such basic facts as who is responsible for the contents of the page or provide a way of verifying that person's credentials for offering information on the topic? How can a Web user know whether to trust information if there is no viable way to determine what influences an advertiser may have on the integrity of that information? How can a Web user know whether to order products from a company if there is no way of verifying that company's legitimacy?

This book discusses these issues and more. It also describes the basic elements that all Web resource creators, new or experienced, need to address when developing online content. By following the suggestions outlined in this book, there is an increased likelihood that a Web author's message will be more successfully conveyed to the Web user.

THE NEED FOR WEB-SPECIFIC EVALUATION CRITERIA

Today's media send out a steady stream of messages intended to entertain, inform, and influence the public's actions and opinions. Understandably, the World Wide Web adds yet another dimension to this daily barrage of messages. Based on a lifetime's exposure to media messages, we develop a set of criteria that we use to evaluate the messages received. Fortunately, the evaluative criteria that we apply to traditional media messages can also serve as a useful starting point for developing methods for evaluating Internet-based resources. Five specific universal criteria—accuracy, authority, objectivity, currency, and coverage—play an essential role in the evaluation process of media content regardless of how it is conveyed.

In addition, several other factors help guide the evaluation process. These include standards and guidelines, regulations, and our own sensory perception. Many information providers adhere to a well-established set of industry standards and conventions regarding the contents and presentation of their materials. Information providers are also obliged to comply with various governmental regulations that affect the content and format of their messages. Using visual and textual cues, an individual can usually differentiate between advertising and informational content in a magazine or newspaper. Similar distinctions occur in radio and television as well. For example, a television commercial is ordinarily distinguishable from the program itself owing to a variety of audio and visual cues. Even an infomercial, a program-length advertisement, is by law accompanied by a disclaimer proclaiming it as a "paid program."

Of course, all of these waters can, and frequently do, get muddied. Whenever a company or organization advertises in a print or broadcast medium, for example, the potential always exists for the contents to be influenced in some manner by the advertiser. Most savvy consumers understand this situation and judge the trustworthiness of the information accordingly.

However, since the Web is a relatively new medium, many standards, conventions, and regulations commonly found in traditional media are largely absent. Lacking many of these traditional formalities, a number of resources have been developed to help Web users locate quality Web information, such as the following:

- Individuals and organizations provide qualitative reviews of Web resources or list resources they have found valuable.
- Experts in various subjects often share lists of quality Web sites relevant to their areas of expertise.
- Academic departments of universities and librarians create pages of authoritative links on topics of interest to their students or patrons.
- News organizations often supply links to Web sites that provide more in-depth information about subjects that they cover.
- A number of health organizations evaluate medical-related sites.

Nonetheless, as valuable as these efforts to review individual sites are, they cannot begin to cover more than a small fraction of the resources available on the Web.

Moreover, although individuals and review services may purport to suggest Web resources on the basis of quality, in reality a site may be listed merely because it has paid money or provided some other type of reward to the reviewer. Therefore, it is still imperative that Web users know how to independently judge the quality of information they find on the Web.

WHAT THIS BOOK INCLUDES

Web resource evaluation strategies are introduced in Chapter 2, with an overview of five traditional evaluation criteria: (1) authority, (2) accuracy, (3) currency, (4) coverage, and (5) objectivity. Chapter 3 discusses the more complex evaluation questions necessitated by characteristics unique to the Web—features such as the use of hypertext links and frames as well as the need for specific software to access certain materials. Chapter 4 examines several new popular Web-based social media tools, namely, weblogs ("blogs") and wikis. The chapter also addresses the unique evaluation challenges associated with each of these tools.

Chapter 5 explores advertising and sponsorship on the Web. It addresses such issues as determining the sponsorship of information content on a Web page and the possible influence an advertiser or sponsor may have on the objectivity of any information provided on the page.

Chapter 6 explores the concepts and issues introduced in the preceding chapters in more detail. It also presents a checklist of basic questions to ask when evaluating or creating any type of Web resource. The chapter also includes annotated screen captures of actual Web pages that illustrate many of the concepts discussed.

Chapters 7 through 12 present an analysis of different types of Web pages based on the framework established in the first section of the book. However, no "one-size-fits-all" approach is adequate for analyzing the diverse array of Web pages. Therefore, Web pages are categorized into the following six types based on their purpose: advocacy, business, informational, news, personal, and entertainment. For example, a business Web page that advertises a company and its products has somewhat different goals from an advocacy Web page created by a political party that urges voters to support a specific legislative initiative. Likewise, a news-oriented page is significantly different from a personal page created by an individual who merely wants to share photos of the family's pets. Therefore, in addition to the checklist of basic questions found in Chapter 6, the book also includes checklists of additional questions to ask when evaluating or creating each specific type of Web page. Each chapter also illustrates the concepts discussed via numerous annotated screen captures.

Chapter 13, the concluding chapter of the book, focuses on Web resource creation issues such as

- Consistent use of navigational aids
- Meta tags
- Basic copyright considerations
- Testing the functionality of a completed Web page or other Web-based resource

A NOTE ABOUT DESIGN ISSUES

Two important aspects of Web resource design are the following:

- Visual design, which consists of aesthetic factors such as the use of images and color.
- Functional design, which consists of factors such as conformity of layout and use of hypertext links to aid in page navigation.

Visual design issues, although important, are well covered in other books and thus are not addressed in this work. However, functional design issues are addressed since they have a significant impact on information quality.

HOW TO USE THIS BOOK

Chapters 2 through 6 are intended to be read consecutively because they serve as the conceptual foundation for the evaluation criteria and the questions that appear in checklists used throughout the second half of the book.

Chapters 7 through 12 are intended to serve as a resource for understanding the six different types of Web pages and the additional questions that need to be asked when either evaluating or creating each type of page. Consequently, these chapters can be either read in consecutive order to gain an understanding of the different types of pages or consulted individually when evaluating or creating a specific type of page.

Although Chapter 13 is designed primarily for individuals who create Web resources, much of the information covered, including that concerning meta tags and copyright, can be useful to both Web users and Web authors.

For the reader's convenience, a complete set of all checklists that appear throughout the book is provided in Appendix A.

To help provide continuity throughout the book, a unique identifier, consisting of a combination of letters and numbers, has been assigned to each important concept introduced in the book. The unique identifier appears each time the concept is repeated in any checklist or illustrated on a screen capture. For example, when the concept of currency is discussed, the following question is asked: Is the date the resource was first placed on the server included somewhere on the page? This question has been assigned the unique identifier CUR 1.2. All identifiers associated with the concept of currency begin with CUR. The number 1.2 following CUR refers to the specific aspect of currency discussed, namely, the date the page was first placed on the server. In addition, whenever this specific concept is illustrated on a screen capture, the identifier CUR 1.2 will appear. Each of the major concepts discussed is denoted with a similar combination of letters and numbers.

The unique identifiers are intended to help the reader readily follow the concepts as they are explained and illustrated. Appendix B contains a complete listing of all the questions accompanied by their unique identifiers.

TWO IMPORTANT CAVEATS

This book presents a variety of techniques for analyzing and presenting Web-based information. Nevertheless, it must be noted that it is possible to follow the techniques outlined in this book to create Web pages and sites that outwardly appear to be trustworthy yet in reality are quite the opposite. This situation obviously creates a dilemma for a Web user attempting to evaluate such resources. The Web, perhaps more than any other medium, inherently possesses these dangers; therefore, regardless of the evaluation techniques employed, there cannot be any absolute guarantees that information that seems to satisfy the evaluation criteria will always be accurate and trustworthy.

Moreover, *Web Wisdom* is not meant to be used as a tool to judge whether a Web resource is "good" or "bad." In fact, without knowing the purpose for which information is intended to be used, this judgment cannot be made. Instead, this book seeks to provide Web users with a method to help them think critically about the Web information they locate and to make their own judgments about whether the information is suitable for their needs.

As previously stated, whether the information is suitable depends on the user's purpose for accessing the information. There may be occasions when certain criteria, such as the need for indicating an author's qualifications to write about a topic, will not be important to the user. For example, if a user has sufficient expertise in a subject area to judge the information quality of a Web resource directly, the resource may be of value even without a listing of the author's credentials. Moreover, if someone is merely seeking opinions on a favorite television show, the absence of an author's name and qualifications may not be critical.

However, in many situations, it *is* important to try to ascertain whether Web information is accurate, authoritative, and reliable. Because of this, it is hoped that both Web users and Web authors will find the tools and techniques presented in this book of value.

DEFINITIONS OF KEY TERMS

Because Web terminology is not always intuitively clear and because certain key concepts are not always defined in a similar way, it is necessary to clarify how the following terms are used throughout the book. It should also be noted that a comprehensive glossary of Web-related terms is provided in Appendix C.

- *Home page:* The page at a Web site that serves as the starting point from which other pages at the site can be accessed. A home page is the Web equivalent to the table of contents of a book.
- *HTML (Hypertext Markup Language):* A set of codes that are used to create a Web page. The codes control the structure and appearance of the page when it is viewed by a Web browser. They are also used to create hypertext links to other pages.
- *Hypertext link ("link"):* A region of a Web page that, once selected, causes a different Web page or a different part of the same Web page to be displayed. A link can consist of a word or phrase of text or an image. The inclusion of

hypertext links on a Web page allows users to move easily from one Web page to another.

- *Search engine:* A tool that can search for words or phrases on a large number of World Wide Web pages.
- *Social networking sites:* "Web sites that allow users to build online profiles; share information, including personal information, photographs, blog entries (see definition below), and music clips; and connect with other users" (U.S. Federal Trade Commission, et al. n.d.).
- *Uniform resource locator (URL):* A World Wide Web address composed of several parts, including the protocol, the server where the "resource" (e.g., a Web page) resides, the path, and the file name of the resource.
- *Web page:* An HTML file that has a unique URL address on the World Wide Web.
- *Web site:* A collection of related Web pages interconnected by hypertext links. Each Web site usually has a home page that provides a table of contents to the rest of the pages at the site.
- *Web subsite:* A site on the World Wide Web that is nested within the larger Web site of a parent organization. The parent organization often has publishing responsibility for the subsite, and the URL for the subsite is usually based on the parent site's URL.
- *Weblog* (also known as a *blog*)*:* A Web page that functions as a publicly accessible unedited online journal. The journal can be formal or informal in nature (U.S. Department of State n.d.; U.S. Legal Services Corporation 2007).
- *Wiki:* A Web site that includes the collaboration of work from many different authors. Also, it is common to allow anyone to edit, delete, or modify the content of a wiki (U.S. Legal Services Corporation 2007).
- *XML (eXtensible Markup Language):* "A *metalanguage*—a language for describing other languages—which lets" Web resource authors create customized markup languages for specific types of documents (U.S. Federal Financial Institutions Examination Council n.d.).

2 Information Quality Criteria for Web Resources

INTRODUCTION

Since the World Wide Web represents a unique combination of conventional and new media, evaluation and creation of Web-based resources require the application of an equally novel mix of long-established and innovative principles. Moreover, Web authors can help establish the quality of their offerings by following some simple guidelines for presenting information online.

A COMPARISON BETWEEN TWO WEB PAGES PRESENTING INFORMATION

Figures 2.1 and 2.2 are both Web pages that might be retrieved using a Web search engine. Both pages have important messages to convey, yet there are striking differences in how effectively these messages are presented. Figure 2.1 shows a section from the Web page with the title The Multinational Corporation (MNC) and Globalization. Although the information appears to be valid, there is no simple way to determine the information's attribution and reliability for the following reasons:

- No author is given for the work, and there is no link to a home page that might identify the author and the author's qualifications for writing on the subject. As a result, we have no way of knowing whether the author is a scholar in the field or a student writing a term paper.
- Without knowing the author's rationale for writing this work, we cannot adequately determine whether the material is intended to be presented in an objective manner, or whether it has been slanted by someone with a particular point of view.
- This page has become separated from the rest of the work, and there are no links to enable a reader to easily locate the other parts. As a result, we cannot determine what other topics are included in the work and to what depth these topics are addressed.
- Brief citations are provided for the factual information included on the page. However, since the page has become separated from its bibliography, we cannot access the full citations, which would likely be needed to retrieve the original works and validate the facts presented.

URL provides no obvious clues about the origin of the page

Author's name not provided and no link to a home page listing:
• the author's name
• his or her qualifications
• the purpose for writing the piece

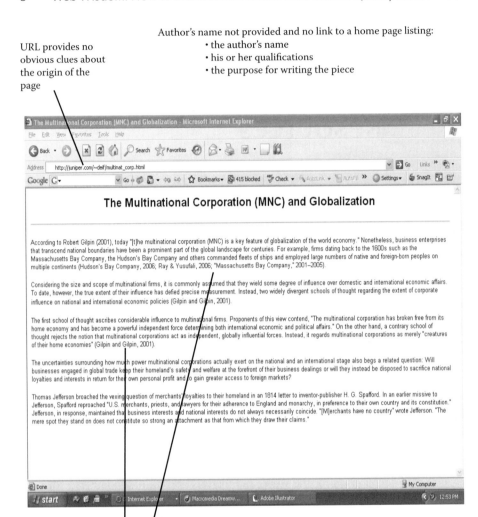

Citations for factual information are given; however, there is no link to a bibliography listing the information needed to access the cited works

FIGURE 2.1   A Web page, The Multinational Corporation (MNC) and Globalization. (Web page by author.)

In contrast, Figure 2.2, the page titled The American Summer Colony at Cobourg, Ontario, provides us with the following information that we can use to help determine its authorship and reliability:

• The page clearly indicates who is responsible for the information.
• Contact information for the page's author is provided on the page.
• The purpose of the page is described.

Clear statement of project's goals

The American Summer Colony at Cobourg, Ontario

In the years following the Civil War, Cobourg Ontario, a community nestled on the shores of Lake Ontario, emerged as one of the most popular resort communities in North America. Families from the southern United States found a summer respite in a cool and hospitable climate without needing to spend their money in the northern United States. Somewhat surprisingly, at the same time, wealthy families from the northern United States--including Pennsylvania--also began summer pilgrimages to Cobourg. The northerners were attracted to the Cobourg area's reputedly high quality ozone and for various business-related reasons. A number of the seasonal Cobourg residents later constructed large mansions throughout the town and its environs. Indeed, Cobourg would remain a popular destination for both southern and northern families alike until the early years of the 20th century.

Overview and Goals of the Project

Drawing upon an array of primary and secondary sources, this project examines Cobourg's summer populace during the period 1865-1930s. Specifically, the project addresses the following questions: 1) What specific factors contributed to Cobourg's popularity with both southern and northern families? 2) What types, if any, of economic and social relationships existed between northern and southern families living in Cobourg? 3) What types of economic and social relationships existed between the American families and Cobourg's permanent Canadian residents?; 4) What factors contributed to Cobourg's loss of favor as a summer residence for the Americans?

Despite Cobourg's popularity with wealthy Americans, it faced major competition from other resorts in Ontario , especially in the Lake Muskoka region. Like Cobourg, the Lake Muskoka region boasted a large U.S. summer colony which included many prominent families from Pennsylvania and elsewhere. Consequently, the Cobourg project has been recently expanded to examine the rival Lake Muskoka summer colony and its similarities/differences to its Cobourg counterpart.

Bibliographic information for all primary and secondary sources as well as other resources related to Cobourg's American summer colony and its denizens--and to a lesser extent Lake Muskoka's colony and its denizens--located during the project are being made publicly available on this site. The site is intended to serve as a research portal for individuals in the United States, Canada, and throughout the world who are interested in learning more about the topic. Additional resources will be placed on the site as the project continues. On a final note, while every attempt is made to ensure the accuracy of the information provided, errors may still occur. If you believe any information provided on this site is incorrect, please contact the site's administrator, Marsha Ann Tate.

Cobourg, Ontario: Canada's Mason - Dixon Community (Text of Presentation Given at the Cobourg & District Historical Society's 25th Annual Dinner, May 24, 2005.) (PDF format)	Cobourg, Ontario: Canada's Mason - Dixon Community (PowerPoint Presentation in PDF format)	Cobourg Yesterday & Today (Photo Gallery)
Selected List of Cobourg's U.S. Summer Residents (Arranged Alphabetically by Last Name) (PDF format)	Selected List of Lake Muskoka's U.S. Summer Residents (Arranged Alphabetically by Last Name) (PDF format; Last updated February 4, 2008).	Additional Resources Related to Cobourg and the Town's Summer Residents (PDF format)

I would like to thank the Cobourg & District Historical Society, the Archives of Ontario, the Pennsylvania Canadian Studies Consortium, and the citizens of Cobourg for their assistance with this project.

Page created and maintained by Marsha Ann Tate, PhD
Librarian & Website Coordinator
Department of Plant Pathology
222 Buckhout Laboratory
The Pennsylvania State University
University Park, PA 16802
Phone: 814-865-7736
Questions or Comments? mat1@psu.edu

Return to Marsha's Home Page
Return to Penn State's Home Page
Page created May 18, 2005. Last updated February 4, 2008.

Contact information provided
(postal and e-mail addresses,
phone number)

Link to home page

FIGURE 2.2 A Web page, The American Summer Colony at Cobourg, Ontario. (Web page by author.)

- There is a link to the home page of the individual responsible for the page.
- A date on the page indicates the currency of the information.
- Organizations that have provided funding for or other assistance with the project are clearly indicated.

Although Web users may not be familiar with the page's author, the page provides enough evidence to help them determine whether the information on it is likely to be trustworthy.

Both of these pages convey what appears to be valuable information, yet there is a great disparity between them with respect to verifying the quality of the information provided. This chapter discusses the factors that must be addressed to present information that can be identified as reliable and authoritative. Understanding these same factors will also aid Web users in determining whether the information they reference is coming from reliable, trustworthy sources.

FIVE TRADITIONAL EVALUATION CRITERIA AND THEIR APPLICATION TO WEB RESOURCES

This section describes five traditional evaluation criteria—authority, accuracy, objectivity, currency, and coverage/intended audience. These criteria have their origins in the world of print media but are universal criteria that need to be addressed regardless of the medium evaluated. To provide a more in-depth understanding of the criteria, each is addressed individually. Moreover, since significant overlap often occurs between criteria, these scenarios are also discussed in detail. For example, authority and accuracy often go hand in hand and thus may need to be considered together to achieve a more complete picture of a particular resource.

AUTHORITY

Authority is the extent to which material is the creation of a person or organization recognized as having definitive knowledge of a given subject area.

Authority of Traditional Sources

There are several methods to assess the authority of a work. One approach is to determine an author's qualifications for writing on the subject by examining his or her background, experience, and formal credentials related to the subject area.

Another method for assessing the authority of a work is to examine the publisher's reputation. A publisher earns a reputation for the quality of its materials based on numerous factors, such as the following:

- The accuracy of the contents of its publications
- The types of individuals who use the company's publications
- Reviews written about the publisher's works
- The expertise of the authors writing for the publisher

A publisher that wants to produce quality works must establish and adhere to strict editorial and ethical standards that emphasize quality. The publisher employs editors and ombudsmen (i.e., individuals who hear and investigate complaints against the publication) who continually monitor the information presented. If these practices are consistently and effectively employed, the publisher should gain a reputation for producing publications of excellence and integrity. For example, the publisher of the *Encyclopedia Britannica* has gained a reputation for producing high-quality works largely by following these practices.

Authority of Web Sources

One of the factors that have contributed to the explosive popularity of the Web is the ease with which almost anyone can become a Web publisher. Countless people can now easily circumvent the traditional publishing process and communicate their messages directly to a worldwide audience. While this factor is one of the Web's great strengths, it also presents unique evaluation challenges.

On the Web, obtaining sufficient evidence to adequately evaluate a work can prove quite difficult. For example, as demonstrated in Figure 2.1, there is no guarantee that the author's name or qualifications will be provided. Also, even if an author's name is given on a page, it should not be automatically assumed that this person is the actual author. Moreover, it is often difficult to verify who, if anyone, has ultimate responsibility for publishing the material.

ACCURACY

Accuracy is the extent to which information is reliable and free from errors.

Accuracy of Traditional Sources

Traditional media utilize a number of checks and balances to help ensure the accuracy of content. These include the following:

- The use of editors and fact-checkers to monitor accuracy.
- The use of the peer review process to monitor the accuracy of scholarly journal articles.
- The use of style manuals to help maintain uniformity in language usage and manuscript format.
- The listing of sources for factual information, as appropriate.

Evaluation of information encompasses a large part of our daily lives, albeit we are often not consciously aware of the process. Even a simple trip to the supermarket requires making a large array of evaluation decisions. We commonly compare products on the basis of such objective and subjective criteria as ingredients, prices, calories per serving, size, color, and even shelf location and package appearance. Frequently, our past experience with a particular brand name also plays a major role in our purchasing decisions. For example, if we purchased XYZ brand spaghetti sauce in the past and found it to be flavorful and of overall high quality, we will probably be more likely to purchase the same sauce in the future. Moreover, if faced with a choice between another XYZ brand product and an unfamiliar brand name, we will probably be more apt to favor XYZ brand. In effect, XYZ's spaghetti sauce has earned a good reputation in our view.

We even evaluate information while we watch television. Again, reputation plays a role in the evaluation. However, in this instance, our focus is on the broadcaster's reputation for authority, accuracy, objectivity, and so forth. As a result, we tend to give more credence to information provided on C-Span rather than information offered by an infomercial. As these examples illustrate, reputation often influences our differentiation between the quality of a wide array of products. Consequently, reputation and related factors are revisited several time throughout this book.

As mentioned earlier, authority and accuracy are often interrelated. We often make the assumption that a publisher with a reputation for reliability will produce works that are also accurate. *Consumer Reports*, for example, is a publication found in countless libraries and homes because it has a reputation as an authoritative, reliable source of impartial information. Although readers may not know that the Consumers Union, the publisher of *Consumer Reports*, does not accept any type of funding from the makers of products it tests, they do assume, because of the publication's reputation, that information found in it will be accurate (Consumers Union 1998–2009).

Accuracy of Web Sources

As stated previously, one of the benefits of the Web is that people can easily share their works with the public, independent of traditional publishing or broadcasting venues. Another major advantage of the Web is its timeliness, as Web material can be published almost instantaneously. Nonetheless, several steps used to substantiate the accuracy of traditional media are frequently condensed or even eliminated when works are published on the Web.

This condensation of the traditional publishing process can result in problems as straightforward as the omission of a listing of sources used in research or as complex as what happened in late May 2007 when a television station in Tulsa, Oklahoma, erroneously posted a report of a fire at a Oklahoma refinery on its Web site. Although the station withdrew the story after the refinery categorically denied its authenticity, in the meantime, the posted report triggered a 40-cent increase in U.S. crude prices. In this example, the source of the information—a CBS affiliate—was authoritative, but the Web publishing process had somehow circumvented the checks and balances usually in place to ensure accuracy ("Web Site Error" 2007).

OBJECTIVITY

Objectivity is the extent to which material expresses facts or information without distortion by personal feelings or other biases.

Objectivity of Traditional Sources

No presentation of information can ever be considered totally free of bias because everyone has a motive for conveying a message. However, it is often important to attempt to assess the information provider's objectivity. Knowing the intent of the organization or person for providing the information can shed light on any biases that might be present in the material. For example, we would easily be able to evaluate the objectivity of information originating from the U.S. surgeon general or a tobacco company. Nevertheless, it can be extremely difficult to uncover the biases of information sources with which we are unfamiliar, even print sources, unless the provider explicitly states his or her point of view.

Objectivity of Web Sources

If we are familiar with the author or provider of information on the Web, evaluating its objectivity is probably no more difficult than evaluating the objectivity of print information. However, because the Web so easily offers the opportunity for persons or groups of any size to present their point of view, it frequently functions as a virtual soapbox. It can be difficult, in this jumble of virtual soapboxes, to determine the objectivity of many Web resources unless the purpose of the individual or group presenting the information is clearly stated.

When discussing objectivity, another important factor to consider is the potential influence exerted by advertisers or sponsors on the informational content of works. Although the extent of this influence is difficult to ascertain even in non-Web sources, it has become even more complex on the Web. Because of its complexity, this issue is discussed in greater detail in Chapter 5.

CURRENCY

Currency is the extent to which material can be identified as up to date.

Currency of Traditional Sources

To evaluate the currency of a print source, it is important to know when the material was first created. This information can usually be determined from the publication and copyright dates that commonly appear on the title page or other front matter of a work. However, specific kinds of material may also require additional date-related information beyond these dates. For example, for statistical information, it is important to know not only the publication date but also the date the original statistics were compiled. For example the publication date for the *Statistical Abstract of the United States* may be 2009, but a closer analysis of the contents may reveal the information in many of the charts was collected several years prior to publication. Therefore, reputable print publications that present statistical information also frequently indicate the date the statistics were collected.

Currency of Web Sources

Because there are no established guidelines for including dates on Web pages, it can be difficult to determine the currency of Web resources. Frequently, dates of publication are not included on Web pages, and if included, a date may be variously interpreted as the date when the material was first created, when it was placed on the Web, or when the Web page was last revised.

One advantage of Web publishing is the ease with which material can be revised. However, unless each revision is clearly dated, it can be difficult to keep track of the various versions. This is obviously important if a print or electronic copy has been made of the page for use in research. In addition, because there is no standard format for how dates appear on Web pages, Web users may construe dates differently. Confusion can result when people use different conventions to convey the same information.

COVERAGE AND INTENDED AUDIENCE

Coverage is the range of topics included in a work and the depth to which those topics are addressed. *Intended audience* is the group of people for whom the material was created.

Coverage and Intended Audience of Traditional Sources

Print sources frequently include a preface or introduction at the beginning of the publication explaining the topics the work includes, the depth or level to which these topics are addressed, and the intended audience for the material. If this explanatory material is not included, a table of contents or an index may provide similar information. Even if lacking all of these features, a print source can usually be scanned or browsed to determine the coverage of information and the audience for whom it is written.

Coverage and Intended Audience of Web Sources

Because Internet-based resources often lack the Web equivalent of a preface or introduction, the coverage of the material is often not readily apparent. Moreover, "thumbing" through Web pages can prove to be a tedious process; an index of the site's contents or a site map may be the only practical ways to determine the range of topics and the depth to which they are covered on a particular site.

Likewise, unlike motion pictures and television programs, the majority of Web sources lack rating systems that indicate their intended audience. Thus, the intended audience for the source may only be learned by scanning through its content.

CONCLUSION

The five basic evaluation criteria outlined in this chapter provide a starting point for crafting an evaluation scheme that addresses the "something old, something new" nature of the World Wide Web and its vast array of resources. Chapter 3 focuses on the something new aspects of the Web and the evaluation challenges related to these distinctive features.

3 Additional Challenges Presented by Web Resources

INTRODUCTION

The Web is a hybrid communications channel that integrates many components of traditional media. Like print media, it facilitates the integration of visual content with text. Like film and television, the Web is capable of combining sound and video content. Moreover, other components have been added to this already eclectic media mix. For example, hypertext links facilitate user interaction with the medium by allowing users to make choices concerning how and in what sequence they access Web-based resources. This merging of text, images, motion, sound, and interactive links constitutes a powerful means of communication. Not surprisingly, this potent hybrid medium can, at times, pose complex evaluation challenges. Two of these evaluation challenges relate to advertising, namely: (1) the blending of information and advertising, and (2) the blending of information, advertising, and entertainment. Although both of these advertising, related also exist in conventional media, they can prove even more challenging in a Web-based media environment. Accordingly, Chapter 5 is devoted to these issues.

Some demanding evaluation challenges posed by the Web, however, are not found in traditional media. These unique Web-related challenges include

- The use of hypertext links
- The use of frames
- Search engines that retrieve pages out of context
- Software requirements that limit access to information
- The instability of Web pages
- The susceptibility of Web pages to alteration

Furthermore, over the past few years, yet another group of distinctive online evaluation challenges has emerged thanks to the ever-growing popularity of weblogs, wikis, and many other Internet-based applications and tools collectively known as *social media*. Chapter 4 discusses several of these applications and their associated evaluation challenges.

THE USE OF HYPERTEXT LINKS

The ability to use hypertext to link a variety of pages is one of the Web's most appealing features. However, the fact that one Web page contains material of high information quality does not guarantee that pages linked to the original page will be uniform in quality. As a result, each Web page, and often sections therein, must be evaluated independently for the quality of the information it contains.

THE USE OF FRAMES

Information presented on Web pages within frames can also present an evaluation challenge. A *frame* is a Web feature that allows the division of a user's browser window into several regions, each of which contains a different HTML (Hypertext Markup Language) page. The boundaries between frames may be visible or invisible. Sometimes, each frame can be changed individually, and sometimes one frame in the browser window remains constant while the other frames can be changed by the user.

The contents of the various frames often originate from the same site. Nonetheless, it is possible for the different frames to originate from different sites without the user being aware of it. Consequently, a Web user needs to be alert to the fact that, because the contents of each frame may be originating from a different Web site, each frame needs to be evaluated independently.

DYNAMIC WEB CONTENT

DATABASE-DRIVEN WEB SITES

When a Web site is created using traditional Web authoring techniques, the contents of the pages within the sites remain fixed or "static" until revisions are made to their underlying HTML coding. Likewise, the URLs for the pages remain unchanged until the pages are either moved to another location within the site or transferred to another site or server.

Today, however, static Web pages and URLs are becoming less common as content management systems are increasingly used to create and manage the content on many Web sites. Databases are integral components of content management systems and thud serve as the underlying foundation upon which "database-driven" sites are built. In this new generation of Web sites, Web pages often simply serve as templates for displaying the results of database queries rather than functioning as storage areas for information. Google™, Yahoo!™, and countless other Web sites are constructed around this database-driven model.

Dynamic URLs represent another unique feature of database-driven Web sites. Each time a Web user types a query into a database-driven site, a new "dynamic URL" is generated. Dynamic URLs routinely include characters such as *?, &, $, +, =, %, .cgi,* and *.cgi-bin* (WebMediaBrands 2009a, 2009b). For example, when the phrase "web evaluation" was searched on Yahoo!, the dynamic URL *http://search.yahoo.com/search?p=%2B%22Web+evaluation%22&fr=yfp-t-151&toggle=1&cop=mss&ei=UTF-8* was generated for the search results page.

As the Yahoo example above demonstrates, dynamic URLs can be extremely long and unwieldy, especially if the URL needs to be cited in a paper or publication. Moreover, the fact that a database supplies most of the information displayed on the pages within a database-driven site presents sundry evaluation challenges such as determining the frequency and extent of updates of the information provided.

REALLY SIMPLE SYNDICATION (RSS)

Really Simple Syndication (RSS) represents yet another popular form of dynamic Web content. RSS represents "a family of web formats used to publish frequently updated digital content." Although RSS feeds are typically text-based, they "may also include audio files (podcasts) or even video files (vodcasts)" (U.S. National Oceanic and Atmospheric Administration, National Weather Service n.d.).

A *feed reader*, also known as a *news reader* or *news aggregator*, is an application needed to collect and view RSS content. There are many types of feed readers including "desktop, Web, mail-client, browser plug-in," and more. The readers share a common function namely, to simultaneously "monitor any number of sites and sources while providing near real-time updates from one location" (Library of Congress, undated)

Once a Web user selects and installs a feed reader, the user can subscribe to whatever RSS feeds are of personal interest. A standard icon is used to indicate where RSS feeds are available on a particular Web site; however, the subscription process for feeds varies according to the type of feed reader application used.

A diverse assortment of government agencies, businesses, organizations, and even individuals now offer RSS feeds. For example, Figure 3.1 illustrates the various RSS feed subscriptions available from the whitehouse.gov Web site.

The ability of feed readers to seamlessly monitor updates from a multiplicity of Web sites affords Web users a substantial savings of time and energy. Feed readers are also of value to Web authors since they can be used to automatically aggregate and integrate content from other Web sources into authors' own pages and sites (Library of Congress, undated; U.S. National Archives and Records Administration 2008).

SOFTWARE REQUIREMENTS AND OTHER FACTORS THAT LIMIT ACCESS TO INFORMATION

Beyond the need for a Web user to use a feed reader to view RSS feeds, two additional software-related factors may further limit the user's ability to access all of the information offered on a Web page: (1) the types of browser used, and (2) other supplementary software that may be required to utilize the content.

Different browsers display information in varying ways. As a result, material created to be viewed by one graphical browser may not appear in the same manner when it is viewed by a different one. Moreover, older versions of a browser may display Web content or otherwise function differently from newer versions.

Beyond variations in browsers, other software or hardware may also be necessary to access the full contents of a page or site. A Web site may require a sound card and the appropriate software plug-ins to access multimedia content on the site.

FIGURE 3.1   A Web page listing RSS feeds available at the whitehouse.gov Web site. (Reprinted from United States, The White House, Subscribe to RSS, The White House, Washington, DC, n.d., http://www.whitehouse.gov/rss/ [accessed April 2, 2009].)

Moreover, many forms and other publications on Web sites are exclusively available in Portable Document Format (PDF). Access to these materials requires downloading Adobe Acrobat reader or other software capable of viewing PDF files. Therefore, it is important to realize that these along with other factors may limit access to Web-based resources.

PAGES RETRIEVED OUT OF CONTEXT BY SEARCH ENGINES

Another Web-specific issue involves the retrieval of orphan Web pages by search engines. Most Web sites are designed with the expectation that a user will initially view a page containing background information such as that provided on a home page. Sometimes, however, users will enter the site at another page instead of the home page, as when they retrieve a page by using a search engine. In these instances, there may be no way to determine who is responsible for the page (and other important details) unless this information is provided either on the page itself or on a page linked to it. The Multinational Corporation and Globalization Web page example discussed in Chapter 2 illustrates this problem since the page does not provide a link to the site's home page or include any identifying information. Although it is not always possible to evaluate the authority of such a page, some techniques that can help with this task are outlined in Chapter 6.

THE SUSCEPTIBILITY OF WEB PAGES TO ALTERATION

Web pages are also susceptible to alteration, both accidental and deliberate. Accidental alteration can occur when converting information into a Web-friendly format. For example, text and images that appear correctly in a word-processing document or spreadsheet may be distorted when converted into another format. Also, problems associated with the transmission of data across the Web and other sundry factors can cause odd characters to appear on the page or prevent the entire page from loading.

Deliberate alteration, on the other hand, can result when hackers break into a site and purposely change the information. Given the susceptibility of Web information to alteration, it is always important to compare facts found in a Web-based source with those found in other Web and non-Web sources to verify their accuracy.

THE REDIRECTION OF URLS TO DIFFERENT WEB SITES AND OTHER MALICIOUS ACTIVITIES

In addition to deliberate Web page alteration, Web users must also be alert to another deceptive practice, namely, the redirection of URLs to unwanted or counterfeit pages and sites. Redirection can take several forms. It can be caused by a browser hijacker, a type of spyware that infects a Web user's browser and then changes the user's designated browser home page, delivers pop-up ads on the screen, or automatically redirects the browser to other Web pages and sites (Harvey et al. 2007; U.S. Federal Trade Commission et al. n.d.). Alternately, a Web user may click on a seemingly legitimate hypertext link provided in an e-mail message or on a Web page that, in turn, sends the user to a counterfeit page or site. Unfortunately for Web users, fake sites are becoming ever more sophisticated and often look virtually identical to their legitimate counterparts. Once at a counterfeit site, unsuspecting visitors are often asked to provide personal or financial information to "verify" their account or registration, fill out an "order form," or perform other tasks. In addition, these faux sites may serve as a means for transmitting viruses and other malware to visitors' computers. Moreover, it is possible for redirection to take place even when a Web user types in a legitimate URL address rather than clicking on a hypertext link.

THE INSTABILITY OF WEB PAGES

The Web is inherently a less-stable medium than print. Pages and sites appear and disappear; URL addresses change. Given the dynamic nature of the Web, the contents of a particular page or the entire site itself may no longer be available when a user attempts to revisit it.

Unfortunately, there is relatively little Web users can do about this situation except to be aware of it and, when using the Web for research, to keep track of the URL addresses of the pages visited and make electronic or print copies of important pages.

Web content creators can also take steps to help minimize the difficulties related to the volatility of the Web. Several of these techniques are addressed in later chapters of this book.

CONCLUSION

As outlined in this chapter, the unique features of the World Wide Web have both positive and negative implications for Web evaluation. Acquiring a basic understanding of these features and recognizing how they can be used for malicious purposes will help Web users minimize the potential pitfalls associated with them.

4 Weblogs and Wikis: Social Media Content

INTRODUCTION

Today, weblogs, wikis, and various other social networking tools are seemingly indispensable fixtures of modern-day society. The ubiquitous nature of social media combined with the media's unique characteristics highlights the need for Web users to recognize how these characteristics may influence information derived from these sources. Accordingly, this chapter is devoted to social media and its unique evaluation challenges. The chapter begins with an overview of social media followed by a brief discussion about two popular types of social media tools and services: weblogs and wikis. The chapter concludes with a discussion of several specific evaluation challenges presented by social media.

SOCIAL MEDIA: AN OVERVIEW

The term *social media* refers to a wide variety of Internet and mobile networking applications, such as weblogs (blogs), wikis, microblogs, and more. Social media applications generally share a number of common characteristics: "(a) interactivity, (b) collaboration, (c) aggregation, (d) incremental content, and (e) content replication" (U.S. National Archives and Records Administration 2008). These applications allow families, friends, business associates, and the general public to readily communicate and share information, interests, and opinions with other members of their personal networks or to broader audiences.

Social media usage has surged over the past few years thanks to the popularity of sites such as MySpace, Facebook, Twitter, YouTube, Wikipedia, and various other sites and services. Social media embodies the ongoing static-to-dynamic evolution of Web pages and sites. Also similar to the World Wide Web overall, social media rcprcscnts an amalgam of traditional and new media elements; likewise, social media content can be created by professionals or everyday individuals. Indeed, social media affords individuals an opportunity to share their self-produced media content with a global audience, a practice previously reserved almost exclusively to large multinational media corporations.

In fact, social media content is becoming increasingly integrated with its traditional media counterparts as television networks, newspapers, and radio stations incorporate ever larger amounts of social media content and applications into their programming, Web sites, and other ancillary activities.

For example, viewers of the Cable News Network (CNN) are encouraged to submit their own photos and videos of newsworthy events to CNN's iReport.com Web site. The photos and videos submitted to iReport.com are then made available for public viewing on the site. In addition, selected "iReports" are later featured on CNN's cable television news-related programs. A notice displayed across the top of the iReport.com home page informs visitors that stories submitted to the site "are not edited, fact-checked or screened" with the exception of stories that have been "vetted" aired on CNN. These stories are accordingly marked "ON CNN". (Cable News Network Inc. n.d.).

The next section examines weblogs and wikis, two commonly used social networking applications.

WEBLOGS (BLOGS)

Weblogs, frequently referred to simply as *blogs*, are one of today's favored forms of online communication. In a nutshell, a Weblog refers to a Web site that functions as an unedited online journal for the *blogger* (author). The blogger's periodic journal entries, which usually appear in reverse chronological order, are variously known as *blogposts*, *weblog posts*, *postings*, or merely *posts*. Many bloggers also make it possible for readers to post comments about their blogposts on the site. Finally, bloggers and blogposts are collectively referred to as the *blogosphere* (Technorati 2008; U.S. National Archives and Records Administration 2008; U.S. Legal Services Corporation 2007).

A blog can be formal or informal in nature; consist solely of text posts; or alternately incorporate photos, video clips, or RSS (Really Simple Syndication) feeds. The specific content and character of any individual blog varies according to a number of factors, including, among others, the personality and technical proficiency of the blog's creator and the amount of time the creator can devote to blogging. Figure 4.1 shows an example of a blog created as a supplementary resource for *Web Wisdom*.

Initially regarded as just another online venue for individuals to share their life experiences, opinions, hobbies, or pastimes, blogging has rapidly grown in popularity as its communicative powers are more fully appreciated by the wider society. The expanding power and prestige of blogs is reflected by the number of businesses, organizations, and governmental bodies that have jumped aboard the "blog bandwagon" over the past few years.

WIKIS

Wikis are another widely used type of social media application. A *wiki* is defined as "a Web site that includes the collaboration of work from many different authors." Each wiki posting "is versioned so that postings can be compared." In addition, "all past entries are kept in a log as a version of the evolving discussion." Wikis can be used for a variety of tasks, including (a) collaborative writing, (b) collaborative projects, (c) "finding consensus around an issue or concept (e.g., virtual meetings), and (d) vocabulary development" (U.S. National Archives and Records Administration 2008). Like weblogs, wikis are being used by a diverse array of groups and organizations. For example, Figure 4.2 illustrates the home page for the FHA Wiki, a wiki created by the U.S. Department of Housing and Urban Development, Federal Housing Administration. The FHA Wiki provides definitions of home financing-related terms and provides information about the FHA's programs and services. This particular wiki is not currently editable by the general public. Figure 4.3 shows an example of an entry from the FHA Wiki.

FIGURE 4.1 A weblog. (Reprinted from Marsha Ann Tate blog, Web Wisdom: How to Create Information Quality on the Web, 2008–2009, https://blogs.psu.edu/mt4/mt.cgi [accessed April 2, 2009].)

EVALUATION CHALLENGES PRESENTED BY SOCIAL MEDIA CONTENT

Unfortunately, the inherent strengths of social media, namely, its immediacy, interactivity, and capacity to aggregate and "put content in new contextual patterns," also can make evaluating information derived from social media sources a tricky task (U.S. National Archives and Records Administration 2008). Attempting to determine

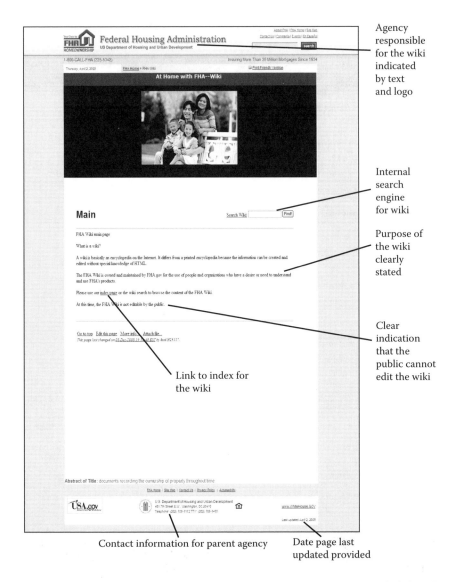

Agency responsible for the wiki indicated by text and logo

Internal search engine for wiki

Purpose of the wiki clearly stated

Clear indication that the public cannot edit the wiki

Link to index for the wiki

Contact information for parent agency

Date page last updated provided

FIGURE 4.2   A wiki home page. (Reprinted from U.S. Federal Housing Administration, 2008-a, FHA Wiki [page last changed December 8, 2008], U.S. Federal Housing Administration, Washington, DC, http://portal.hud.gov/portal/page?_pageid=73,1829262&_dad=portal&_schema=PORTAL [accessed April 2, 2009].)

the authority, accuracy, and objectivity of a blog site or blogpost, wiki site, or wiki entry can prove especially challenging. When evaluating information provided by a blog or wiki, use the following questions to supplement the general questions found on the Checklist of Basic Elements:

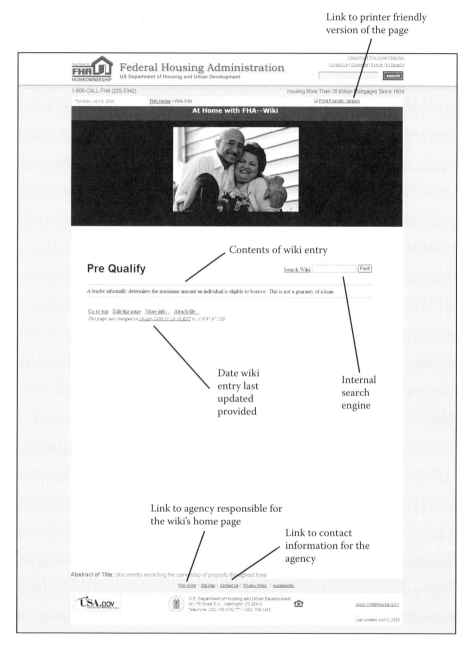

FIGURE 4.3 A wiki entry. (Reprinted from U.S. Federal Housing Administration, 2008-b, Pre qualify [page last changed April 29, 2008], in FHA Wiki, U.S. Federal Housing Administration, Washington, DC, http://portal.hud.gov/portal/page?_pageid=73,1829262&_dad=portal&_schema=PORTAL [accessed April 2, 2009].)

When evaluating information from a blog:

- Who is hosting or sponsoring the blog?
- Does the author of the blog provide links or citations to sources of factual information? If so, attempt to go back to the original sources.
- Does the author's profile provide any information regarding his or her qualifications for writing on the subject?
- Is the blog cited on other Web sites? If so, which ones?
- Is the blog cited in newspapers, television/radio newscasts, or other conventional media outlets? If so, which ones?
- Does the author maintain other blogs, Web pages, or sites?

When evaluating information from a wiki:

- Who is hosting or sponsoring the wiki?
- Who is authorized to add, modify, or delete information on the wiki?
- Are the names of the wiki's contributors listed?
- Are links or citations to sources of factual information provided? If so, attempt to go back to the original sources.
- Does the wiki have an editor or fact-checker?
- Are there earlier versions of the wiki entries? If so, how do they differ from the current version?

Like Web users, Web authors also need to use caution when delving into the social media realm. Web authors who are interested in creating a blog or wiki on a third-party Web site also need to consider the following:

- Who owns the copyright for materials contributed/posted on the site?
- Where are the blogposts and other related information included on the site physically stored?
- Who has access to the content beyond the creator(s) of the blog or the wiki?
- How long will the material remain on the site?
- Does the site owner provide a method for blog or wiki authors to permanently delete content that they have created from the site?
- If personal information is stored on the site, what measures are in place to secure it from unauthorized users?

Finally, it is important for both users and creators of social media to carefully read the terms of service for social media sites, especially if they are planning to contribute content to them.

CONCLUSION

Social media's interactive, collaborative, and aggregative capacity; flexibility; and immediacy make it a formidable communicative force. Given this powerful combination of elements, social media presents a host of novel challenges for Web users and authors alike. These challenges are not insurmountable; however, they may require additional expenditures of time and energy to successfully address them.

5 Advertising and Sponsorship on the Web

ADVERTISING, SPONSORSHIP, AND INFORMATION ON THE WEB

Advertising and sponsorship are hardly new phenomena. They have long been the mainstays of newspapers and television as well as art, music, sporting events, and countless other activities. Advertising and sponsorship have traditionally served as a means for businesses and organizations to promote their products, services, and ideas in return for financial and other support for their activities.

However, the Web has introduced a number of new twists to traditional advertising and sponsorship. The multimedia nature of the Web, in combination with features such as hypertext links, frames, and cookies, has encouraged the formation of a wide array of alliances among advertisers, sponsors, and information providers. Under these circumstances, Web users often face a daunting task when attempting to ascertain the influence an advertiser or sponsor may exert on the information provided on a Web page or site.

The Web's added nuances to advertising and sponsorship have also inspired a new vocabulary. For example, Internet marketing, also referred to as online marketing or E-marketing, simply refers to marketing that "uses the Internet." A related term, interactive marketing, is likewise Internet focused; however, it elevates Internet marketing to a new level at which the marketer engages in a "conversation" with the customer by addressing the customer, being aware of what the customer conveys, and fashioning a response based on the customer's input (Arizona Office of Tourism n.d.).

From the Web user's perspective, the depth of analysis into the potential influences of advertisers and sponsors on information depends mostly on how he or she ultimately intends to use the acquired information. For example, it would certainly be more important to determine the potential influence of an advertiser when seeking medical information or shopping for a new car than when looking for a new television program or movie to view. Therefore, in some cases it may be more crucial than in others to untangle these relationships.

DEFINING ADVERTISING AND SPONSORSHIP

Because advertising and sponsorship play significant roles in our everyday lives, it seems that it would be an easy task to find universally accepted definitions for the terms. Unfortunately, this is not the case; instead, scholars, businesspeople, marketers, and the general public each ascribe somewhat different meanings to advertising and sponsorship. In some instances, the two terms are treated distinctly; in other instances, they are considered virtually interchangeable. For the purposes of this book, *advertising* is defined as the conveyance of persuasive information, frequently by paid announcements and other notices, about products, services, or ideas. Conversely, *sponsorship* is defined as financial or other support given by an individual, business, or organization for something, usually in return for some form of public recognition.

Since these definitions encompass a diverse array of activities, they have been subdivided into the following categories: commercial advertising, advocacy advertising, institutional advertising, word-of-mouth advertising, corporate sponsorship, and nonprofit sponsorship.

COMMERCIAL ADVERTISING

Commercial advertising is "advertising that involves commercial interests rather than advocating a social or political cause" (Richards 1995–1996). It is designed to sell a specific product or service. Usually, the consumer can readily identify the product or service being sold. Commercial advertising can assume a number of forms:

- Ads in print newspapers and magazines.
- Radio and television commercials.
- Billboards.
- Product placement, the visual or verbal reference to a product in another form of communication. For example, companies often pay producers or studios a fee to have their products appear on or be mentioned by a character in a film or television show.
- Endorsements and testimonials.
- Direct mail brochures.
- Web banner and pop-up ads.
- Web pages and sites designed primarily to promote specific products and services.

Figure 5.1 illustrates a common form of online commercial advertising, a home page from a company Web site devoted to promoting the company's products.

ADVOCACY ADVERTISING

Advocacy advertising is advertising that promotes political or social issues. Examples of advocacy advertising include ads promoting the following:

- Public health, such as youth antismoking and AIDS prevention
- Public safety, such as fire prevention or the use of seat belts
- The conservation of natural resources and wildlife, such as limiting the use of carbon-based fuels and protecting endangered species

Government agencies and nonprofit organizations are often sources for advocacy advertising.

FIGURE 5.1 Commercial advertising. (Reprinted from Roots Canada Ltd., Roots Canada & International [home page], 2002–2009-c, http://canada.roots.com/ [accessed March 31, 2009]. Reproduced with permission from Roots Canada Ltd.)

Figure 5.2 is an example of advocacy advertising in the form of a banner advertisement promoting public television that appears on the home page of WPSU, a member-supported public media organization in central Pennsylvania. Clicking on the banner ad takes the user to an advocacy page, We Need You to Advocate for Public Television (n.d.), located at another .org Web site.

INSTITUTIONAL ADVERTISING

Institutional advertising is "advertising to promote an institution or organization rather than a product or service, in order to create public support and goodwill" (Richards 1995–1996). Institutional advertising is meant to convey the idea that the organization enhances the community in some way.

WORD-OF-MOUTH ADVERTISING

Word-of-mouth advertising is the endorsement of a product or service by an individual who has no affiliation with that product or service other than being a user of it and who is not being compensated for the endorsement. Examples of word-of-mouth advertising include a person recommending a product or service to a friend during a conversation or an individual mentioning a product on his or her Web page. The mixing of word-of-mouth advertising and social media has produced a new phenomenon known as *viral advertising* or *viral marketing*. *Viral advertising* is defined as "marketing techniques

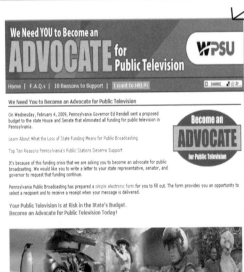

The "ADVOCATE" banner ad on the home page of WPSU, a member supported public media organization links to a Web page at a another site that encourages public advocacy and action on behalf of WPSU and public television in general.

FIGURE 5.2   Advocacy advertising. (Reprinted from Penn State Public Broadcasting, WPSU/Home, 2004–2009, http://www.wpsu.org/ [accessed March 27, 2009]; We need you to become an advocate for public television, n.d., http://www.wqln.org/advocate/default.aspx?sid=wpsu [accessed March 27, 2009]. Reproduced with permission from Penn State Public Broadcasting.)

that use pre-existing social networks to produce increases in brand awareness, through self-replicating viral processes analogous to the spread of pathological and computer viruses." The techniques "facilitate and encourage people to pass along a marketing message voluntary." Viral advertising comes in a variety of forms, including text messages, games, images, and audio or video clips (Arizona Office of Tourism n.d.).

CORPORATE SPONSORSHIP

Corporate sponsorship occurs when a company provides financial or other material support for something, usually in return for some form of public recognition. Sporting and cultural events as well as Web pages and sites are frequently partially or fully supported through corporate sponsorship.

NONPROFIT SPONSORSHIP

Nonprofit sponsorship consists of financial or other material support by an individual or nonprofit organization, usually in return for public recognition.

DISTINGUISHING AMONG ADVERTISING, SPONSORSHIP, AND INFORMATION ON THE WEB

THE OVERLAPPING AND BLENDING OF ADVERTISING AND SPONSORSHIP ON THE WEB

Although the categorization of advertising and sponsorship is a beneficial theoretical exercise, in reality the types as defined here often defy such orderly classification. Different kinds of advertising and sponsorship are often so extensively intertwined that they become almost indistinguishable. Likewise, the concepts of advertising and sponsorship frequently overlap when they are applied to actual Web sites. This is due, in part, to the fact that both advertisers and sponsors are often identified via a banner ad on a Web page. Without a clear explanation of whether the banner ad signifies advertising or sponsorship, it is difficult for a user to differentiate between the two. In addition, the banner ads of both advertisers and sponsors are frequently linked to the Web site of an advertiser or sponsor. Therefore, when a site has a corporate sponsor, it regularly provides a link from its home page to the corporate sponsor's site. These links can facilitate a direct transfer from the announcement of corporate sponsorship on the sponsored Web site to commercial advertising offered at the corporate sponsor's own Web site.

As the Our Thanks page on the Math Forum @ Drexel University Web site illustrates (Figure 5.3), events, projects, and services are frequently jointly sponsored by government, commercial, and nonprofit entities. In this example, the site's collaborators and sponsors include, among others, the National Science Foundation (NSF), an independent U.S. government agency; Shodor, a nonprofit organization; and Texas Instruments, a for-profit corporation.

Like advertising and sponsorship, advertising and information are also regularly blended together on the Web. For example, business Web sites commonly promote products or services while providing a significant amount of seemingly objective

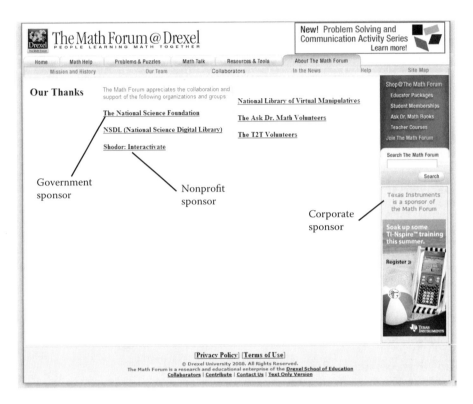

FIGURE 5.3 Combined government, corporate, and nonprofit sponsorship of a Web site. (Reprinted from The Math Forum @ Drexel University, The Math Forum @ Drexel University: Our thanks, 2008, http://mathforum.org/appreciation.html [accessed March 25, 2009]. Reproduced with permission from Drexel University, copyright 2009 by The Math Forum @ Drexel. All rights reserved.)

information related to problems or issues that the product or service promoted on the site is designed to address.

Some medical sites, for example, are sponsored by a physician who ostensibly provides objective information about a specific medical problem. However, the same site may also be promoting the physician's services in the form of a surgical procedure or medication claiming to cure the malady. Since a definite conflict of interest exists in these instances, any information provided on the sites must be viewed accordingly. The critical point to remember is that although Web resources frequently provide helpful free information, the user must always consider what potential factors may influence the objectivity, and thus the trustworthiness, of the information presented.

To more fully understand the complex relationships among sponsorship, advertising, and information on the Web, it is useful to examine how sponsorship and advertising interact with informational content in print publishing.

In traditional print publications, there are usually clear visual distinctions between advertising and editorial, or informational, content. Even when advertising and information are mixed, as in the case of an advertorial that presents a significant amount of information but is, in reality, an advertisement for something, the print convention is to identify the information as an advertisement somewhere on the page. Thus, the phrase "special advertising feature" or some other message alerts readers that what follows is information carefully blended with an advertisement. Smart consumers know to beware of the objectivity of information presented in this manner.

However, on the Web there are few, if any, standards to ensure that a visual distinction exists between advertising and information or that advertorial material is labeled as such. As a result, information on the Web is often seamlessly blended with advertising.

Traditionally, in print publishing there also exists a policy of separation between the advertising and the editorial department (i.e., the department that produces the information content) of a publication. Where the policy is more rigorously followed, the advertising department is not supposed to influence the editorial department. The purpose of this arrangement is to have the information produced by the editorial department as free as possible from advertisers' pressure to bias the information in some manner. In practice, of course, there is great variation in how strongly the policy is adhered to by different publishers.

As stated above, the Web is largely devoid of any established conventions or anything that remotely resembles the separation between advertising and editorial content. One notable exception to this is the American Society of Magazine Editors (ASME), "the professional organization for editors of consumer magazines and business publications, which are edited, published, and sold in the United States" (ASME n.d.-a). ASME has addressed this issue in its editorial guideline, "Best Practices for Digital Media," which provides recommended procedures which are intended to help readers readily delineate between "independent editorial content and paid promotional information" in Internet-based publications (ASME n.d.-b).

However, ASME's efforts to preserve the separation between advertising and editorial content is a rarity on the Web. Consequently, the Web user needs to be constantly vigilant for advertiser influence on the objectivity of information.

The advertising–editorial content demarcation is not even applicable, to a significant percentage of traditional media content. For example, in many print publications the advertising and informational content are produced by the same organization, as is the case with promotional brochures such as those published by a business to advertise its products or services. Nevertheless, in the world of print publishing, readers have learned to recognize most publications of this type. For example, we do not assume that a brochure produced by a car dealership is going to provide unbiased information about the makes and models of vehicles it sells, and we know how to evaluate the material accordingly.

Similarly, when Web sites feature advertising on them, the advertiser and the organization with overall responsibility for the site are frequently the same entity. Therefore, the material at these Web sites has more in common with a car dealership's brochure than with a magazine article published by a company that maintains separate advertising and editorial departments. However, on the Web it is often not so readily apparent when an individual or group is supplying both the informational and advertising content of the page.

A Continuum of Objectivity on the Web

To better understand the potential effects of the blending of advertising, sponsorship, and information on the Web, it is helpful to view Web sites on a continuum ranging from sites that accept no outside advertising to sites designed exclusively for marketing a company's own products and services. Many Web sites fall somewhere between the two extremes of this continuum since they incorporate various forms of advertising and sponsorship into their online content. In these instances, the influence of advertisers and sponsors on the objectivity of the information provided varies widely. Consequently, whenever any site not only accepts advertising and sponsorship but also provides information, the user must be aware of the potential influence by the advertiser or sponsor on the objectivity of that information.

Hypertext Links and the Blending of Advertising, Information, and Entertainment

Hypertext links help facilitate the blending of advertising, information, and entertainment content on the Web. For example:

- Outside advertisers are highly motivated by marketing concerns to place a link on a site, which, once followed, attracts customers from the original site to their own site.
- Some business sites lure people to the company-sponsored site by providing links to free entertainment. Once at the site, marketing to these people can be readily achieved.
- As a way of attracting people, business sites also may offer a listing of links to information perceived to be objective. Again, the intention is that these visitors will respond to the company's advertising while using the site.

Affiliate marketing represents one popular form of online advertising. This type of marketing involves the placement of a hypertext link, icon, or other kind of advertisement publicizing a business or organization on a Web site owned by another individual, business, or organization (the "affiliate"). Each time a visitor, subscriber, or customer "clicks through" the affiliate's site to reach and subsequently purchase something at the business's site, the affiliate will receive a monetary or other reward from the business in return (Arizona Office of Tourism n.d.). Amazon.com is one well-known online company that engages in affiliate marketing (as illustrated in Figure 5.4). The Math Forum @ Drexel University, participates in the affiliate marketing programs of Amazon.com as well as Target, a discount chain store.

SORTING OUT THE RELATIONSHIP BETWEEN ADVERTISERS, SPONSORS, AND INFORMATION

Because linkages among advertising, sponsorship, and information on the Web are commonly more labyrinthine than those in traditional media, Web users need to become adept at sorting out these interrelationships. They must learn to identify

Example of affiliate marketing

FIGURE 5.4 Affiliate marketing. (Reprinted from The Math Forum @ Drexel University, The Math Forum @ Drexel University [home page], 2009, http://mathforum.org/index.html [accessed April 3, 2009]. Reproduced with permission from Drexel University, copyright 2009 by The Math Forum @ Drexel. All rights reserved.)

the key stakeholders and analyze as best as possible what influence they might have on each other and on the objectivity of the information provided. The following are some useful questions to ask about advertisers and sponsors found on a Web site's pages. If the information on the page is being provided for free:

- What seems to be the purpose of the information provider for making the information available? Is the purpose one that might influence the objectivity of the information?
- What are some of the possible influences on the objectivity of the information at the site? For example, what are the potential influences of commercial advertisers, corporate sponsors, or the author of the information?

In addition to knowing whether the information on the page is provided for free, it is also important to know what kind of organization is providing the information. Some types of organizations that provide information on the Web include the following:

Advocacy Groups. When an advocacy group offers information on the Web for free, users should assume the information will be biased in a certain direction to support the organization's goals. Even if the organization provides information from a reputable journal or other outside source, users can assume they will not find both sides of the issue represented.

Nonprofit Organizations. Even when information appears to be provided by a source such as a nonprofit organization, users must be aware of potential conflicts of interest that might arise. For example, when a piece of research is presented on a nonprofit hospital's Web site, a corporate sponsor such as a drug company may have directly supported the research. If this is the case, the hospital needs to make this relationship clear so that the reader can understand that there may be a possible conflict of interest.

Commercial Businesses. When a business offers information on its own Web site, the questions that need to be asked are somewhat different. Some information, such as software documentation and product pricing, will be objective. However, users must not assume that all the information will be objective because the company's goal is to promote its own products and services. Therefore, readers need to ask the following questions:

- What is the company's purpose for offering the information?
- How are the products the company is promoting related to the information being provided?
- Are there offers for free or discounted products in exchange for some type of information from the user?
- If free entertainment is provided, what relation does it have to the products or services being offered?
- Is the business withholding more detailed information that is only available for a fee or requires registration to access it?
- Is marketing information being gathered; if so, for what purpose?

STRATEGIES FOR ANALYZING WEB INFORMATION PROVIDED BY SITES THAT HAVE ADVERTISERS OR SPONSORS

The following are three strategies that can be helpful when sorting out the relations among advertisers, sponsors, and information on the Web:

1. *Identify the key stakeholders involved in providing information at the site.* It is important to identify who is involved in providing the information at a

Web site. Is the site a subsite of a larger organization? If so, what is its relationship to this organization? Does the site appear to have any corporate, government, or nonprofit sponsors?

2. *Identify what information at the site is in actuality advertising.* An analysis of what appears to be objective information at a business site may reveal that the information is biased in favor of the company. Figure 5.5 is an example of a site that provides a link to a supposedly objective bibliography about evaluating Web resources. However, a closer examination of the page reveals that one of the entries on this supposedly objective list of resources is in reality a link to a page designed to promote the company's products and services. By placing this link in the midst of links to legitimate evaluation sources, the company hopes to confer legitimacy on its own site.

3. *Identify the purpose for providing entertainment at the site.* Some business sites provide entertainment as a way of drawing in users so that they can be given a marketing message. The site shown in Figure 5.5 blends entertainment and advertising in this way.

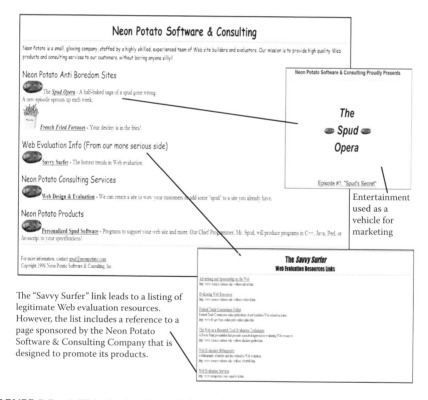

FIGURE 5.5 A Web site that blends information, advertising, and entertainment. (Brenda Corman, Ken Robinson, and Marsha Ann Tate, fictitious Web site, 1998–2009.)

CONCLUSION

This chapter has presented concepts that are important to understand when trying to sort out the relationships among advertisers, sponsors, and Web information. The chapter has also stressed the importance of understanding the influence these relations might have on the objectivity of the information. However, just because a site includes advertising does not necessarily mean that the information contained at the site is not objective. Similarly, an absence of advertising does not guarantee that the material at the site is without bias. Therefore, when assessing the trustworthiness of a site, it is not enough just to determine the site's advertisers and sponsors. It is also important to assess the trustworthiness and authority of the person, organization, or business responsible for the information at the site. The next seven chapters include numerous tools and techniques to aid a Web user in analyzing the potential trustworthiness of information found at a Web site.

6 Applying Basic Evaluation Criteria to a Web Page

HOW TO USE CHAPTERS 6 THROUGH 12

This chapter introduces basic elements important to include on any Web page. The chapter also identifies six different types of Web pages (advocacy, business, informational, news, personal, and entertainment) and discusses the need to include additional elements on each of these types of pages.

When using Chapters 6 through 12 to evaluate or create Web pages:

1. Read Chapter 6 to learn the basic elements that need to be included on any page, regardless of its type. Also, review the Checklist of Basic Elements located at the end of the chapter.
2. Determine what type of page you are evaluating or creating. Each of the following six chapters is devoted to one of the six types of pages and begins with information to assist in identifying that type of page.
3. Consult the appropriate chapter to learn what elements, in addition to the basic ones listed in this chapter, need to be included when evaluating or creating this specific type of page.

INCORPORATION OF THE BASIC ELEMENTS INTO WEB PAGES

This section discusses the key Web page elements and provides illustrations of how they are incorporated into actual Web pages. These elements are the following:

- Authority (site level)
- Authority (page level)
- Accuracy
- Objectivity
- Currency
- Coverage and intended audience

AUTHORITY (ELEMENTS 1 AND 2)

Authority is the extent to which material is the creation of a person or group recognized as having definitive knowledge of a given subject area. When discussing the authority of information on the Web, it is first helpful to analyze the authority of the overall Web site and then to analyze the authority of a specific page within the site.

Element 1: Authority (Site Level)

One crucial aspect of evaluating a Web site as a whole is ascertaining the authority of the site. Figure 6.1 is an illustration of the home page from the Canadian Media Companies at Home and Abroad Research Project Portal. From an analysis of this home page, we can determine the following factors:

- The individual responsible for the page is affiliated with the Pennsylvania State University and a link is provided to the university's home page.
- A link is provided to the author's personal home page, which, in turn, provides a link to the author's vita.
- The address, phone number, and e-mail address for the individual responsible for the page is provided.
- The goals and objectives of the project are provided on the page.

Questions to Ask

The following questions are important to consider when analyzing the authority of a Web site. The items referred to in the following questions should be located either on the page itself or on a page directly linked to it:

- Is it clear what organization, company, or person is responsible for the contents of the site? This can be indicated via a logo. Without this basic information, it is virtually impossible to verify the authority of the site.
- If the site is a subsite of a larger site created by an individual, company, or organization, does the subsite provide the name of the person or group that has ultimate responsibility for the contents of the subsite and can help add legitimacy to the information provided at the subsite?
- Is there a way to contact the individual, company, or organization responsible for the contents of the site? These contacts can be used to help verify the legitimacy of the site. Methods of contact may include a phone number, mailing address, or e-mail address. A mailing address can be especially helpful in determining the legitimacy of a particular site.
- Are the qualifications of the individual, company, or organization responsible for the contents of the site indicated? Including such qualifications is crucial if the site does not originate from a familiar source.
- If all of the site's contents are protected by a single copyright holder, is the name of the copyright holder given? The copyright holder is often the same as the contact point, but if not, it can provide additional information about the authority of the site.
- Does the site list any recommendations or ratings from outside sources?

Element 2: Authority (Page Level)

To evaluate the authority of a Web page, you must look at both the authority of the page itself and the authority of the site on which the page resides. Before looking at the authority of the page, whenever possible it is helpful first to return to the site's home page to analyze the authority of the site.

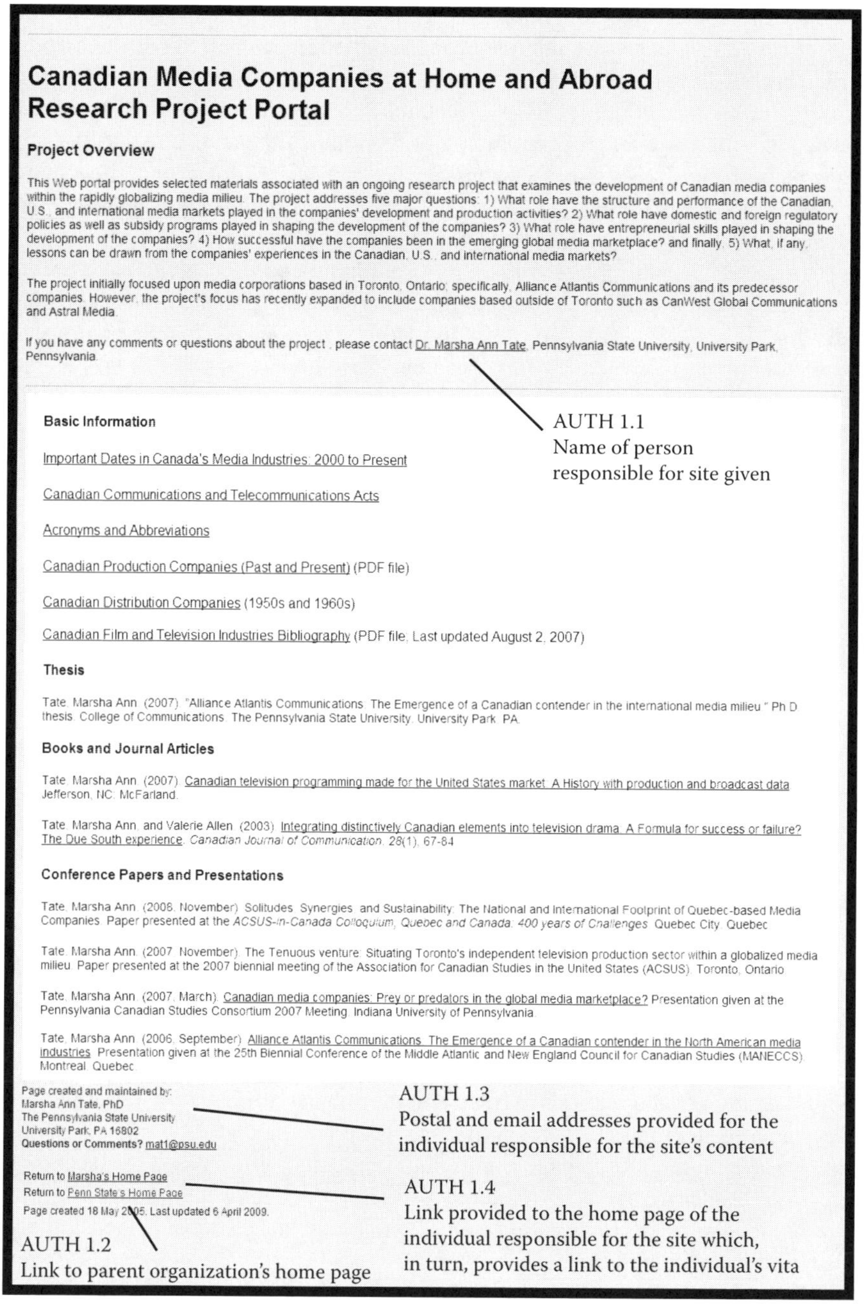

Canadian Media Companies at Home and Abroad Research Project Portal

Project Overview

This Web portal provides selected materials associated with an ongoing research project that examines the development of Canadian media companies within the rapidly globalizing media milieu. The project addresses five major questions: 1) What role have the structure and performance of the Canadian, U.S., and international media markets played in the companies' development and production activities? 2) What role have domestic and foreign regulatory policies as well as subsidy programs played in shaping the development of the companies? 3) What role have entrepreneurial skills played in shaping the development of the companies? 4) How successful have the companies been in the emerging global media marketplace? and finally, 5) What, if any, lessons can be drawn from the companies' experiences in the Canadian, U.S., and international media markets?

The project initially focused upon media corporations based in Toronto, Ontario; specifically, Alliance Atlantis Communications and its predecessor companies. However, the project's focus has recently expanded to include companies based outside of Toronto such as CanWest Global Communications and Astral Media.

If you have any comments or questions about the project, please contact <u>Dr. Marsha Ann Tate</u>, Pennsylvania State University, University Park, Pennsylvania.

Basic Information

AUTH 1.1
Name of person
responsible for site given

<u>Important Dates in Canada's Media Industries: 2000 to Present</u>

<u>Canadian Communications and Telecommunications Acts</u>

<u>Acronyms and Abbreviations</u>

<u>Canadian Production Companies (Past and Present)</u> (PDF file)

<u>Canadian Distribution Companies</u> (1950s and 1960s)

<u>Canadian Film and Television Industries Bibliography</u> (PDF file; Last updated August 2, 2007)

Thesis

Tate, Marsha Ann. (2007). "Alliance Atlantis Communications: The Emergence of a Canadian contender in the international media milieu." Ph.D. thesis. College of Communications. The Pennsylvania State University. University Park, PA.

Books and Journal Articles

Tate, Marsha Ann. (2007). <u>Canadian television programming made for the United States market: A History with production and broadcast data</u>. Jefferson, NC: McFarland.

Tate, Marsha Ann. and Valerie Allen. (2003). <u>Integrating distinctively Canadian elements into television drama: A Formula for success or failure?</u> The Due South experience. Canadian Journal of Communication. 28(1). 67-84

Conference Papers and Presentations

Tate, Marsha Ann. (2006, November). Solitudes, Synergies, and Sustainability: The National and International Footprint of Quebec-based Media Companies. Paper presented at the ACSUS-in-Canada Colloquium, Quebec and Canada: 400 years of Challenges. Quebec City, Quebec.

Tate, Marsha Ann. (2007, November). The Tenuous venture: Situating Toronto's independent television production sector within a globalized media milieu. Paper presented at the 2007 biennial meeting of the Association for Canadian Studies in the United States (ACSUS). Toronto, Ontario.

Tate, Marsha Ann. (2007, March). <u>Canadian media companies: Prey or predators in the global media marketplace?</u> Presentation given at the Pennsylvania Canadian Studies Consortium 2007 Meeting. Indiana University of Pennsylvania.

Tate, Marsha Ann. (2006, September). <u>Alliance Atlantis Communications: The Emergence of a Canadian contender in the North American media industries.</u> Presentation given at the 25th Biennial Conference of the Middle Atlantic and New England Council for Canadian Studies (MANECCS). Montreal, Quebec.

Page created and maintained by:
Marsha Ann Tate, PhD
The Pennsylvania State University
University Park, PA 16802
Questions or Comments? <u>mat1@psu.edu</u>

AUTH 1.3
Postal and email addresses provided for the
individual responsible for the site's content

Return to <u>Marsha's Home Page</u>
Return to <u>Penn State's Home Page</u>
Page created 18 May 2005. Last updated 6 April 2009.

AUTH 1.4
Link provided to the home page of the
individual responsible for the site, which,
in turn, provides a link to the individual's vita

AUTH 1.2
Link to parent organization's home page

FIGURE 6.1 Keys to verifying authority (site level). (Web site created by author.)

Figure 6.2, Food Safety and Imports: An Analysis of FDA Import Refusal Reports, is an example of a page that provides considerable information about the authority of the persons and organization offering the information. The words "United States Department of Agriculture [USDA], Economic Research Service," along with the USDA logo appear together prominently at the top of the page. Links to the home pages of both the Economic Research Service and its parent agency, the USDA, are also provided on the page. E-mail contacts are provided for the agency responsible for the contents of the page as well as the site's Web master.

The individuals responsible for the informational contents of the page are also clearly indicated. Using the Economic Research Service's internal search engine, a query for "Jean C. Buzby," one of the individuals responsible for the page's informational content, returns a list of other materials she has written as well as more information about her qualifications for writing on the topic. Another method to help verify an author's qualifications is to perform a search on Google or other general search engines to see what additional information about the author can be located.

Questions to Ask

The following questions are important to consider when ascertaining the authority of an individual Web page. The items referred to in these questions should be located either on the page itself or on a page directly linked to it.

- Is it clear what organization, company, or person is responsible for the contents of the page? For a page written by an individual with no organizational affiliation, it is important to indicate responsibility for the page.
- If the material on the page is written by an individual, are the author's name and qualifications for providing the information clearly stated? Even though the author of the page has the official approval of that institution, listing the author's qualifications for writing the material gives the page added authority.
- Is there a way of contacting the author? That is, does the person list a phone number, mailing address, or e-mail address? These contact points can be an important way of verifying that an individual is who he or she claims to be.
- Is there a way of verifying the author's qualifications? That is, is there an indication of the author's expertise in the subject area or a listing of memberships in professional organizations related to the topic?
- If the material on the page is copyright protected, is the name of the copyright holder given? As with the copyright holder for a site, the copyright holder for the page is another indication of who has ultimate responsibility for the contents of the page.
- Does the page have the official approval of the company, organization, or person responsible for the site? For pages at a government or business site, as in the USDA Economic Research Service example in Figure 6.2, consistent

AUTH 1.2
Name and logo of parent agency
responsible for contents of site

AUTH 1.1
Name of agency responsible for
contents of site

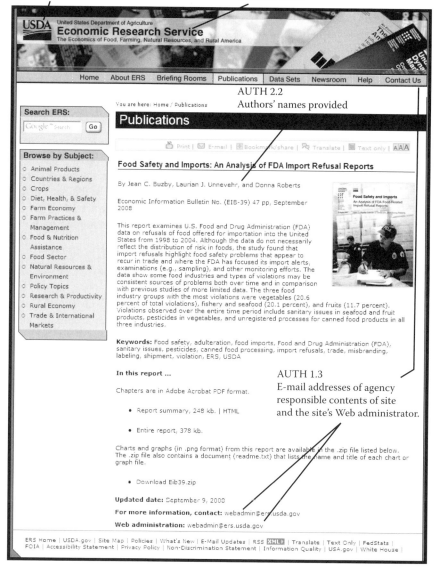

FIGURE 6.2 Keys to verifying authority (page level). (Reprinted from U.S. Department of Agriculture [USDA], Economic Research Service, Food safety and imports: An analysis of FDA food-related import refusal reports, USDA Economic Research Service, Washington, DC, updated September 9, 2008, http://www.ers.usda.gov/Publications/EIB39/ [accessed April 3, 2009].)

page layout and design features often indicate that the page has the official approval of the government agency or business. If the page does not have such official approval, there will often be a disclaimer stating so. There may be times, however, when a Web page has been retrieved by a search engine and there is no indication on the page of who has responsibility for it. In such a situation, the following strategies may assist a user in determining the source of the information:

- If possible, return to the home page because it is usually one of the best sources for discovering what person or organization is responsible for the contents of the page.
- Analyze the URL (uniform resource locator) of the page to see if it offers clues regarding who is responsible for the information on the page.
- Attempt to truncate the URL address by removing the end of it to determine if the page that contains the original link to the page being evaluated can be retrieved. For example, if the page's URL is http://www.host.com/~jsmith/cc17.htm, delete *cc17.htm* and attempt to go to the linking page.

ELEMENT 3: ACCURACY OF THE INFORMATION

Accuracy is the extent to which information is reliable and free from errors. The page Secondhand Smoke: Questions and Answers (Figure 6.3) includes several useful indicators to help determine the accuracy of the information provided. First, the page is free of spelling and typographical errors. This fact does not ensure the accuracy of the contents, but pages free of errors in spelling, punctuation, and grammar do indicate that care has been taken in producing the pages. Second, readers can independently verify the factual information included on the pages. Not only are the sources of the factual information named, but links are also given to additional sources of information provided by the National Cancer Institute (NCI). In addition, it is clear that the NCI has ultimate responsibility for the accuracy of the information provided.

Questions to Ask

The following are important questions to consider when determining the accuracy of a Web page:

- Is the information free of grammatical, spelling, and typographical errors? As stated earlier, these types of errors not only indicate a lack of quality control but also can actually produce inaccuracies in information.
- Are sources for factual information provided so the facts can be verified in the original source? A user needs to both verify that authoritative sources have been used to research the topic and also be able to access the sources cited, if desired.
- If there are any graphs, charts, or tables, are they clearly labeled and easy to read? Legibility is a critical element to consider when converting graphs, charts, or tables into electronic form.

ACC 1.5
Indication of who has ultimate responsibility
for the accuracy of the material

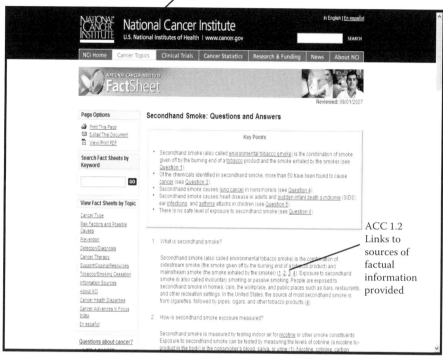

ACC 1.2
Links to
sources of
factual
information
provided

ACC 1.1
Page is free of spelling and typographical errors

FIGURE 6.3 Keys to verifying the accuracy of a Web page. (Reprinted from U.S. National Institutes of Health, National Cancer Institute, Second hand smoke: Questions and answers, National Cancer Institute, Bethesda, MD, reviewed August 1, 2007, http://www.cancer.gov/cancertopics/factsheet/Tobacco/ETS [accessed April 3, 2009].)

ELEMENT 4: OBJECTIVITY OF THE INFORMATION

Objectivity is the extent to which material expresses facts or information without distortion by personal feelings or other biases.

Because of the possibility of influence by an advertiser or sponsor on the objectivity of the information, it is important first to look at any advertising or sponsorship present. It is also important to analyze to what degree the information provider intends to be objective. This can often be difficult to determine, particularly for individual authors. However, when analyzing an organization's pages, clues can sometimes be obtained by looking at the mission statement. For example, as shown in Figure 6.4, the Math Forum @ Drexel University site includes a page detailing the organization's mission and history. The page also provides links to the Math Forum's newsletter as well as to past proposals and reports it has submitted to the National Science Foundation.

An Online Math Education Community Center

    The Math Forum @ Drexel

Awards || Background || Timeline || Connecting to Drexel || Staff || Associates

The Math Forum is a leading center for mathematics and mathematics education on the Internet. Operating under Drexel's School of Education, our mission is to provide resources, materials, activities, person-to-person interactions, and educational products and services that enrich and support teaching and learning in an increasingly technological world.

OBJ 1.9 Mission statement lists goals of organization

Our online community includes teachers, students, researchers, parents, educators, and citizens at all levels who have an interest in math and math education. We work together toward this end in the following ways:

Encouraging communication throughout the mathematical community
We are committed to creating discussion opportunities for all those interested in math. Through our growing collection of mailing lists, Web-based discussion areas, and ask-an-expert services, we give you places to talk, to reach others with similar interests, and to find answers to your burning questions. See Discussions, Bridging Research and Practice (BRAP) and the T2T Teachers' Lounge.

Offering model interactive projects
The Forum's volunteer 'math doctors' and our archive of answers will help you with your math questions - see Ask Dr. Math. Questions about teaching and math education may be sent to our Teacher2Teacher service. Our Problems of the Week (PoWs) provide creative, non-routine challenges for students in grades three through twelve. The archived ESCOT Problems provide interactive challenges for middle and high school students.

Making math-related web resources more accessible
Want to use or develop educational technology? Visit Math Tools, the Forum's community digital library supporting the use and development of software for mathematics education. When a generic Web directory falls short of your mathematics needs, visit the Forum Internet Mathematics Library, which covers math and math education Web sites in depth. In our collaboration with the Mathematical Association of America, Mathematical Sciences Digital Library (MathDL), we collect mathematics instructional material with authors' statements and reader reviews; and catalogs mathematics commercial products, complete with editorial reviews, reader ratings and discussion groups. The Problems Library offers a convenient interface for searching and browsing the collective archives of the six Problem of the Week services.

Providing high-quality math and math education content
There's a lot of material on the Web, but how good is it, and how does it take advantage of new technologies or implement new pedagogy? We have worked with teachers, students, and researchers to put the best of their materials on the Web. This collaborative work is available via the Forum's Teacher Exchange: Forum Web Units. Teachers are invited to use the Web interface to contribute their own lessons.

Growing with the Web
We notify you about new sites of interest, feature the latest and best, and host focused and timely discussions of math education and associated source materials. Subscribe to our Newsletter.

Proposals and reports to the National Science Foundation
1996 grant proposal
1999 annual report
JOMA Digital Library Applet project proposal
Math Tools project proposal

OBJ 1.18 Links to information about the organization's sponsors and collaborators

About our services

Ask Dr. Math	Math Tools	Math Forum Internet News
Teacher2Teacher	Internet Mathematics Library	MathDL
Problems of the Week	Teacher Exchange	JOMA

Postal Address	Phone/Fax
The Math Forum	800 756 7823
3210 Cherry Street	215 895 1080
Philadelphia PA 19104	215 895 2964 (fax)

[Privacy Policy] [Terms of Use]

Home || The Math Library || Quick Reference || Search || Help

 The Math Forum is a research and educational enterprise of the Drexel School of Education.

Questions to Ask

The following are questions important to consider when analyzing the objectivity of Web information:

- Is the point of view of the individual or organization responsible for providing the information evident? It is important to know to what degree the information provider is attempting to be objective with the information offered on the Web page.
- If there is an individual author for the material on the page, is the point of view of the author evident?
- If there is an author for the page, is it clear what relationship exists between the author and the person, company, or organization responsible for the content? It is important to know to what extent the entities responsible for the content might influence the objectivity of the author.
- Is the page free of advertising? If not, it is important to try to determine to what extent an advertiser might influence the information provided.
- If there is advertising on the page, is it clear what relationship exists between the company, organization, or person responsible for the informational contents of the page and any advertisers represented on the page?
- If there is both advertising and information on the page, is there a clear differentiation between the two?
- Is there an explanation of the site's policy relating to advertising and sponsorship?
- If the site has nonprofit or corporate sponsors, are they clearly listed? If so, it is important to try to determine to what extent the sponsor might influence the informational content.
- Are links included to the sites of any nonprofit or corporate sponsors where a user can learn more about them?
- Is additional information provided about the nature of the sponsorship, such as an indication of what type it is (nonrestrictive, educational, etc.)?

ELEMENT 5: CURRENCY OF THE INFORMATION

Currency is the extent to which material can be identified as up to date. For example, the currency of an author's personal home page (Figure 6.5) can be easily determined since both the date it was originally posted on the server and the date it was last revised appear on the page. In addition, the dates appear in a format readily understood by international readers.

FIGURE 6.4 (Opposite) Keys to verifying the objectivity of a Web site. (Reprinted from The Math Forum @ Drexel University, About the Math Forum: mission and history, 1994–2009-a, http://mathforum.org/about.forum.html [accessed April 3, 2009]. Reproduced with permission from Drexel University, copyright 2009 by The Math Forum @ Drexel. All rights reserved.)

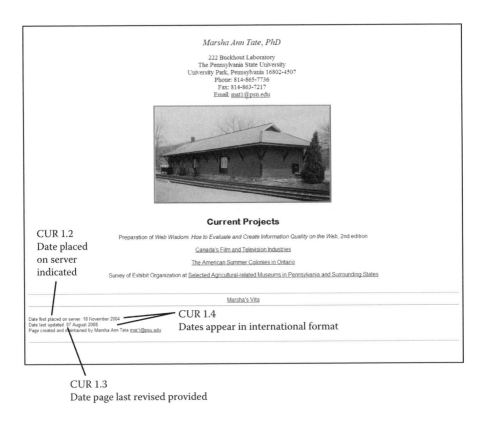

FIGURE 6.5   Keys to verifying the currency of a Web page. (Web page created by author.)

Questions to Ask

The following are questions important to consider when determining the currency of Web information:

- Is the date the material was first created in any format included on the page?
- Is the date the material was first placed on the server included on the page?
- If the material has been revised, is the date (and time, if appropriate) it was last revised included?
- To avoid confusion, are all creation and revision dates provided in an internationally recognized format? Examples of dates in international format (day month year) are 5 January 2009 and 29 October 2012.

ELEMENT 6: COVERAGE OF THE INFORMATION AND ITS INTENDED AUDIENCE

Coverage is the range of topics included in a work and the depth to which those topics are addressed. The *intended audience* is the group of people for whom the material is created. For example, the home page of the Math Forum @ Drexel University Web site (Figure 6.6) provides an overview of the site, a description of the site's intended audience, as well as links to the site's various contents. In addition, the page

COV/IA 1.1
Indication of types of materials found at the site

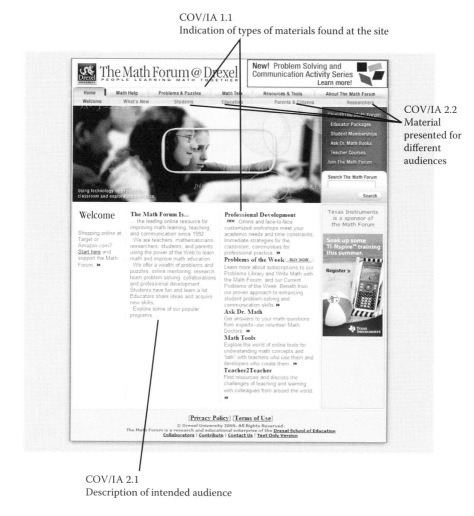

COV/IA 2.2
Material
presented for
different
audiences

COV/IA 2.1
Description of intended audience

FIGURE 6.6   Keys to verifying the coverage and intended audience of a Web site (Reprinted from The Math Forum @ Drexel University, The Math Forum @ Drexel University, 2009. http://mathforum.org/index.html [accessed April 3, 2009]. Reproduced with permission from Drexel University, copyright 2009 by The Math Forum @ Drexel. All rights reserved.)

provides sections designed specifically for students, educators, parents and citizens, and researchers.

Questions to Ask

The following are questions important to consider when determining the coverage and intended audience of a Web site:

- Is it clear what materials are included at the site? This can be difficult to determine unless there is an index to the site or a site map.

- If the site is still under construction, is the expected date of completion indicated?
- If a page incorpartes elements of more than one type of page, is there a clear differentiation between the types of content?
- Is the intended audience for the material clear?
- If material is presented for several different audiences, is the intended audience for each type of material clear?

INTERACTION AND TRANSACTION FEATURES

Interaction and transaction features are an additional category of basic elements important to include on any type of Web page, regardless of its type. Interaction and transaction features are tools that enable a user to interact with the person or organization responsible for a Web site or enter into a transaction (usually financial) via a Web site.

Some ways a user might interact with a site are obvious—for example, filling out an online order form or providing information such as a credit card number. On the other hand, a user may also provide information to a person or organization responsible for a site in less obvious ways that may, in fact, be transparent to the user, as when a site collects information about a visitor via cookies.

Cookies enable data to be stored by a server on a user's computer. Although cookies may expire when a user closes the browser, they are typically stored on the user's computer and can be read by the server that originally supplies them when the user visits the site again.

Some of the features made possible by the use of cookies include the following:

- The storage of user IDs and passwords. Storage of this data eliminates the need for a user to reenter the information each time he/she reenters a particular site.
- The creation of a "shopping cart" into which a user can place items before paying for them.
- The tracking by advertisers of the pages a user visits within a site. This enables advertisers to tailor ads to the user's interests and to monitor the effectiveness of the pages.
- The personalization of a Web page or site according to a user's preferences (Rankin 1998).

Whether a site is collecting information openly by such means as order forms or surreptitiously using mechanisms such as cookies, it is important that users have confidence that the information they are providing to the site will be kept confidential unless the user indicates it may be made public. Therefore, it is important that the site make clear its policy regarding the confidentiality of information collected, both while it is in transit to the site and once it has arrived at the site. This can be done not only by explicitly stating the site's policy on these issues but also by indicating what technical measures the site has in place to ensure privacy.

Other important interaction and transaction features relate to how easily the user can provide feedback to the site and restrictions on the use of the materials offered at the site. Figure 6.7 illustrates the privacy policy and terms of use for the Math Forum @ Drexel University Web site.

The following is a list of important general considerations concerning interaction and transaction features:

- If any financial transactions occur at the site, does the site indicate what measures have been taken to ensure their security? A secure transaction is an encrypted, or scrambled, communication between a Web server and a browser. Because the data communicated in a secure transaction are encrypted, they are less likely to be intercepted by an unauthorized party during the transfer across the Web to its intended destination.
- If the business, organization, or person responsible for the page is requesting information from the user, is there a clear indication of how it will be used?
- If cookies are used at the site, is the user notified? Is there an indication of what the cookies are used for and how long they last?
- Is there a feedback mechanism for a user to comment about the site?
- Are any restrictions regarding downloading and other uses of the materials offered on the page clearly stated?

AN INTRODUCTION TO NAVIGATIONAL AND NONTEXT FEATURES

In addition to the basic elements described above, there are two additional features that need to be considered in the evaluation and creation of Web pages: navigational aids and nontext features. Checklists for these two categories of features are included in Chapter 13, which is devoted exclusively to issues involved in creating effective Web pages and sites. However, this chapter defines these elements and provides examples of how they are used because, although they are critical factors in the creation of Web sites, they also play an important, if more indirect, role in the evaluation of Web information.

CONSISTENT AND EFFECTIVE USE OF NAVIGATIONAL AIDS

Navigational aids are elements that help a user locate information at a Web site and allow the user to move easily from page to page within the site. Navigational aids allow readers the flexibility they need to move to a desired spot in the Web site. They are necessary for two reasons: (1) They allow readers to "browse" easily through the site, and (b) they provide an orientation to the material at the site, just as a table of contents, chapter headings, page numbers, and an index provide an orientation to material within a book.

The following navigational aids are important to include on Web pages and/or sites:

- A browser title
- A page title

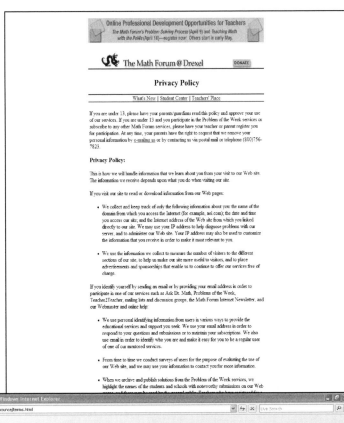

Online Professional Development Opportunities for Teachers
The Math Forum's Problem Solving Process (April 9) and Teaching Math with the PoWs (April 16)—register now! Others start in early May.

The Math Forum @ Drexel   DONATE

Privacy Policy

What's New || Student Center || Teachers' Place

If you are under 13, please have your parents'/guardians read this policy and approve your use of our services. If you are under 13 and you participate in the Problem of the Week services or subscribe to any other Math Forum services, please have your teacher or parent register you for participation. At any time, your parents have the right to request that we remove your personal information by e-mailing us or by contacting us via postal mail or telephone (800)756-7823.

Privacy Policy:

This is how we will handle information that we learn about you from your visit to our Web site. The information we receive depends upon what you do when visiting our site.

If you visit our site to read or download information from our Web pages:

- We collect and keep track of only the following information about you: the name of the domain from which you access the Internet (for example, aol.com); the date and time you access our site; and the Internet address of the Web site from which you linked directly to our site. We may use your IP address to help diagnose problems with our server, and to administer our Web site. Your IP address may also be used to customize the information that you receive in order to make it most relevant to you.

- We use the information we collect to measure the number of visitors to the different sections of our site, to help us make our site more useful to visitors, and to place advertisements and sponsorships that enable us to continue to offer our services free of charge.

If you identify yourself by sending an email or by providing your email address in order to participate in one of our services such as Ask Dr. Math, Problems of the Week, Teacher2Teacher, mailing lists and discussion groups, the Math Forum Internet Newsletter, and our Webmaster and online help:

- We use personal identifying information from users in various ways to provide the educational services and support you seek. We use your email address in order to respond to your questions and submissions or to maintain your subscriptions. We also use email in order to identify who you are and make it easy for you to be a regular user of one of our mentored services.

- From time to time we conduct surveys of users for the purpose of evaluating the use of our Web site, and we may use your information to contact you for more information.

- When we archive and publish solutions from the Problem of the Week services, we highlight the names of the students and schools with noteworthy submissions on our Web

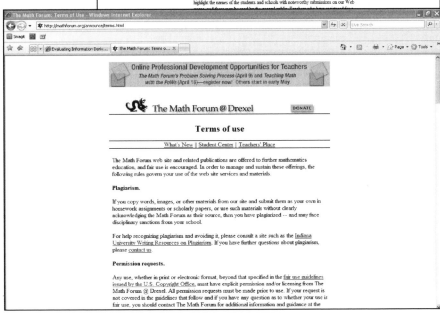

The Math Forum: Terms of Use - Windows Internet Explorer

http://mathforum.org/announce/terms.html

Snagit

Evaluating Information Deriv... | The Math Forum: Terms o...   Page ▾ Tools ▾

Online Professional Development Opportunities for Teachers
The Math Forum's Problem Solving Process (April 9) and Teaching Math with the PoWs (April 16)—register now! Others start in early May.

The Math Forum @ Drexel   DONATE

Terms of use

What's New || Student Center || Teachers' Place

The Math Forum web site and related publications are offered to further mathematics education, and fair use is encouraged. In order to manage and sustain these offerings, the following rules govern your use of the web site services and materials.

Plagiarism.

If you copy words, images, or other materials from our site and submit them as your own in homework assignments or scholarly papers, or use such materials without clearly acknowledging the Math Forum as their source, then you have plagiarized -- and may face disciplinary sanctions from your school.

For help recognizing plagiarism and avoiding it, please consult a site such as the Indiana University Writing Resources on Plagiarism. If you have further questions about plagiarism, please contact us.

Permission requests.

Any use, whether in print or electronic format, beyond that specified in the fair use guidelines issued by the U.S. Copyright Office, must have explicit permission and/or licensing from The Math Forum @ Drexel. All permission requests must be made prior to use. If your request is not covered in the guidelines that follow and if you have any question as to whether your use is fair use, you should contact The Math Forum for additional information and guidance at the

- Internal links within a site
- The URL for the page
- Bread crumbs (show the location of a page within the Web sites hierarchy)
- A site map or index
- An internal search engine for the site (if appropriate)

Browser Title

The *browser title* is the title of a page that is picked up by the browser from the HTML (Hypertext Markup Language) <TITLE> tag. It has the following characteristics:

- It usually appears as part of the browser frame at the top of the browser window (in the "title bar").
- It is distinguished from the page title, which is the title that appears in the body of the Web page.
- The presence of the browser title allows the user to quickly identify the contents of the page.

Browser Title Examples

The following is an example of a home page browser title that clearly indicates both the company responsible for the page and that the page is the home page.

The Neon Potato Company Home Page

The following are examples of non–home page browser titles that clearly indicate both the company responsible for the page and the specific contents of the page:

Neon Potato: Company Information
Neon Potato: Product Information
Neon Potato: Copyright Information

Additional Points about Browser Titles

Whatever appears in the HTML <TITLE> tag will typically become the default title of any browser bookmark to the page. The browser title should be descriptive of the page's contents so it can be easily recognized in a bookmark list. (A bookmark is a URL address stored on a user's computer that allows the user to easily return to a frequently visited page. The ability to store bookmarks is a common browser capability.)

Whatever appears in the HTML <TITLE> tag is usually picked up by a search engine and used as the default description of that page. Therefore, each browser title should be concise, yet descriptive. When creating browser titles, it may be

FIGURE 6.7   (Opposite) The Math Forum @ Drexel University Web site's privacy policy and terms of use. (From The Math Forum @ Drexel University, The Math Forum @ Drexel privacy policy, 1994–2009-b, http://mathforum.org/announce/privacy.html [accessed April 3, 2009]; The Math Forum @ Drexel terms of use, 1994–2009-c, http://mathforum.org/announce/terms.html [accessed April 3, 2009]. Reproduced with permission from Drexel University, copyright 2009 by The Math Forum @ Drexel. All rights reserved.)

helpful to think of the title of a site as the title of a book and the title of a page as a chapter heading of a book.

Page Title

The *page title* is the title found in the text of the Web page. The presence of a page title allows the user to quickly identify the contents of the page. It is often created by using an HTML heading (usually an <H1>) tag. The page title will frequently be the same as the browser title for the page.

URL for the Page

The URL is an identifier that uniquely distinguishes the page from all other World Wide Web pages. Including the URL in the body of the page enables users who print out the page to have a printed record of its source and to revisit the page at a later date.

Hypertext Links

Hypertext links (or simply *links*) are regions of a Web page that, when selected, cause a different Web page or a different part of the same Web page to be displayed. A link can consist of a word or phrase of text or an image. The inclusion of links on a Web page allows users to move easily from one Web page to another.

Including appropriate links allows the user to navigate within the site according to individual preferences, and to return to the site's home page, site map, or for sites arranged in hierarchy, to the page one level up in the hierarchy. Having the ability to return to the site's home page is important in the process of determining the authority of the site, and having the ability to access the site map is important in determining the coverage of the site. In addition, it is helpful to have internal links placed in a consistent place on each page and be uniform in appearance, whether they are created with graphics or text.

Site Map and Index

A *site map* is a display, often graphical, of the major components of a Web site. An *index* is a listing, often alphabetical, of the major components of a Web site. A site map or index provides a quick overview of the pages contained within the entire site, and each can be an important tool in determining the coverage of the site.

Internal Search Engine

In contrast to the well-known Web search engines such as Google and AltaVista that search for words or phrases on a large number of Web sites, an *internal search engine* is one that searches for words or phrases only within one site.

An internal search engine is a helpful navigational aid for sites that present large amounts of information; it allows users to locate information at the site quickly and easily.

Figure 6.8 illustrates how selected navigational aids have been incorporated into The Pennsylvania State University Department of Plant Pathology Web site to improve its functionality.

NAV 1.5
Browser title indicates what site page is from

Bread crumbs provided

NAV 3.1
Links to
University
home page

NAV 3.4
Internal links placed
consistently on each
page

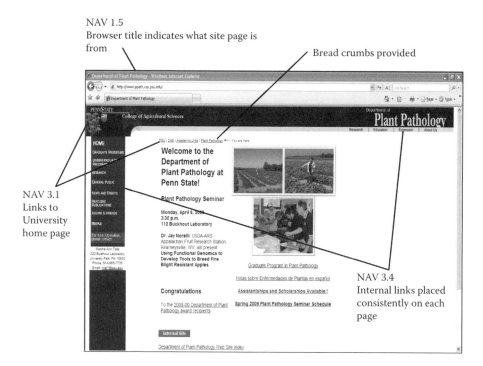

FIGURE 6.8 Examples of navigational aids. (Reprinted from The Pennsylvania State University, Department of Plant Pathology, Department of Plant Pathology [home page], last modified March 27, 2009, http://www.ppath.cas.psu.edu/ [accessed April 3, 2009]. Reproduced with permission from The Pennsylvania State University.)

EFFECTIVE USE OF NONTEXT FEATURES

Nontext features include those elements that require the user to have additional software or a specific browser to utilize the contents of the Web page. Some examples of nontext features include graphics, image maps, sound, and video.

When nontext features are present at a site, some users may not have the ability to take advantage of them. Users may, for example, be viewing the site with the following:

- A text-only browser
- The browser's ability to display graphics turned off
- Special software designed for those who are visually or physically challenged

Therefore, some important considerations for the use of nontext features include the following:

- If the page includes a graphic such as a logo or an image map, is there a text alternative for those viewing the page in text-only mode?
- If the page includes a nontext file (such as a sound or video file) that may require additional software to play it, is there an indication of the additional software needed and where it can be obtained?

- If a file requires additional software to access it, wherever possible is the same information provided in another format that does not require the additional software?
- If a page requires a specific browser or a specific version of a browser, does the page specify what is needed and indicate where it can be obtained?
- When following a link results in the loading of a large graphic, sound, or video file, is information provided to alert the user that this will happen?

INFORMATION ON THE SIX TYPES OF WEB PAGES

The first six elements discussed in this chapter have been compiled into a Checklist of Basic Elements that can be used as a basis for evaluating or creating any Web page, no matter what its type. For a discussion of additional issues that need to be addressed when creating or evaluating advocacy, business, informational, news, personal, and entertainment pages, consult Chapters 7 through 12. For the reader's convenience, all of the book's checklists have been assembled together in Appendix A.

Whether you are evaluating existing Web pages or creating new ones, it is important to analyze them page by page (and, in some cases, section by section within a page) rather than assuming that all pages at a given site will be of one type. For example, it is common to find both informational and advocacy pages at the same site, and also common to find sites that have business pages combined with entertainment pages. Furthermore, personal Web sites often combine different types of pages. It is not uncommon for a personal site to include items about a favorite celebrity, provide information about a favorite research topic, advocate a favorite cause, and try to sell a used bicycle—all at the same time.

Not only can a site contain pages of different types, but also individual pages can be a combination of several different types. Such combination pages may require the use of additional checklists, as appropriate.

THE CHECKLIST OF BASIC ELEMENTS: KEYS TO EVALUATING OR CREATING WEB PAGES

The following questions are general ones that need to be asked when evaluating or creating any Web page, no matter what its type. Answering these questions will help the user determine whether the information on a Web page comes from an authoritative, accurate, and reliable source. The greater the number of "yes" answers, the greater the likelihood that the quality of the information on the page can be determined. The questions can also be used by Web authors as a guide to creating pages that can be recognized as originating from a reliable, trustworthy source.

AUTHORITY (AUTH)

Authority of a Site

The following information should be included either on a site's home page or on a page directly linked to it:

- Is it clear what organization, company, or person is responsible for the contents of the site? This can be indicated by the use of a logo. **AUTH 1.1**
- If the site is a subsite of a larger organization, does the site provide the logo or name of the larger organization? **AUTH 1.2**
- Is there a way to contact the organization, company, or person responsible for the contents of the site? These contact points can be used to verify the legitimacy of the site. Although a phone number, mailing address, and e-mail address are all possible contact points, a mailing address and phone number provide a more reliable way of verifying legitimacy. **AUTH 1.3**
- Are the qualifications of the organization, company, or person responsible for the contents of the site indicated? **AUTH 1.4**
- If all the materials on the site are protected by a single copyright holder, is the name of the copyright holder given? **AUTH 1.5**
- Does the site list any recommendations or ratings from outside sources? **AUTH 1.6**

Authority of a Page

- Is it clear what organization, company, or person is responsible for the contents of the page? Similarity in page layout and design features can help signify responsibility. **AUTH 2.1**

If the material on the page is written by an individual author:

- Is the author's name clearly indicated? **AUTH 2.2**
- Are the author's qualifications for providing the information stated? **AUTH 2.3**
- Is there a way of contacting the author? That is, does the person list a phone number, mailing address, and e-mail address? **AUTH 2.4**
- Is there a way of verifying the author's qualifications? That is, is there an indication of his or her expertise in the subject area or a listing of memberships in professional organizations related to the topic? **AUTH 2.5**
- If the material on the page is copyright protected, is the name of the copyright holder given? **AUTH 2.6**
- Does the page have the official approval of the person, organization, or company responsible for the site? **AUTH 2.7**

ACCURACY (ACC)

- Is the information free of grammatical, spelling, and typographical errors? **ACC 1.1**
- Are sources for factual information provided so that the facts can be verified in the original source? **ACC 1.2**
- If there are any graphs, charts, or tables, are they clearly labeled and easy to read? **ACC 1.4**

Objectivity (OBJ)

- Is the point of view of the individual or organization responsible for providing the information evident? **OBJ 1.1**

If there is an individual author of the material on the page:

- Is the point of view of the author evident? **OBJ 1.2**
- Is it clear what relationship exists between the author and the person, company, or organization responsible for the site? **OBJ 1.3**
- Is the page free of advertising? **OBJ 1.4**

For pages that include advertising:

- Is it clear what relationship exists between the business, organization, or person responsible for the contents of the page and any advertisers represented on the page? **OBJ 1.5**
- If there is both advertising and information on the page, is there a clear differentiation between the two? **OBJ 1.6**
- Is there an explanation of the site's policy relating to advertising and sponsorship? **OBJ 1.7**

For pages that have a nonprofit or corporate sponsor:

- Are the names of any nonprofit or corporate sponsors clearly listed? **OBJ 1.16**
- Are links included to the sites of any nonprofit or corporate sponsors so that a user may find out more information about them? **OBJ 1.17**
- Is additional information provided about the nature of the sponsorship, such as what type it is (nonrestrictive, educational, etc.)? **OBJ 1.18**

Currency (CUR)

- Is the date the material was first created in any format included on the page? **CUR 1.1**
- Is the date the material was first placed on the server included on the page? **CUR 1.2**
- If the contents of the page have been revised, is the date (and time, if appropriate) the material was last revised included on the page? **CUR 1.3**
- To avoid confusion, are all dates in an internationally recognized format? Examples of dates in international format (day month year) are 5 June 2009 and 30 April 2010. **CUR 1.4**

Coverage and Intended Audience (COV/IA)

- Is it clear what materials are included on the site? **COV/IA 1.1**
- If the page is still under construction, is the expected date of completion indicated? **COV/IA 1.2**

- Is the intended audience for the material clear? **COV/IA 2.1**
- If material is presented for several different audiences, is the intended audience for each type of material clear? **COV/IA 2.2**

INTERACTION AND TRANSACTION FEATURES (INT/TRA)

- If any financial transactions occur at the site, does the site indicate what measures have been taken to ensure their security? **INT/TRA 1.1**
- If the business, organization, or person responsible for the page is requesting information from the user, is there a clear indication of how the information will be used? **INT/TRA 1.2**
- If cookies are used at the site, is the user notified? Is there an indication of what the cookies are used for and how long they last? **INT/TRA 1.3**
- Is there a feedback mechanism for users to comment about the site? **INT/TRA 1.5**
- Are any restrictions regarding downloading and other uses of the materials offered on the site clearly stated? **INT/TRA 1.9**

7 Keys to Information Quality in Advocacy Web Pages

KEYS TO RECOGNIZING AN ADVOCACY WEB PAGE

The primary purpose of an advocacy Web page is to influence public opinion by attempting to influence people's views on a topic or encouraging activism. A single individual or group of people may be responsible for the page. The URL address of an advocacy page often ends in .org (organization) if the page is sponsored by a nonprofit organization. Examples of advocacy organizations include the Democratic and Republican parties, the National Right to Life Committee, and the National Abortion Rights Action League.

A "yes" answer to any of the following questions provides a good indication that the primary purpose of the page is advocacy. Does the page:

- Seek to influence people's opinion on something?
- Seek to influence the legislative process?
- Encourage contributions of money?
- Try to influence voters?
- Promote a cause?
- Attempt to increase membership in an organization?
- Provide a point of contact for like-minded people?

ANALYSIS OF ADVOCACY WEB PAGES

As noted in Chapter 4, government agencies frequently create or sponsor advocacy Web pages and sites, as illustrated in Figures 7.1 and 7.2. Figure 7.1 shows the home page of the EPA Environmental Kids Club, a Web site sponsored by the U.S. Environmental Protection Agency (EPA). The youth-oriented site promotes environmental awareness and education. On the other hand, Figure 7.2 shows Can Your Food Do That, a page from Smallstep Kids, a Web site created by the U.S. Department of Health and Human Services to encourage youth to adopt healthy eating and lifestyle habits. Figures 7.1 and 7.2 show some of the elements that are important to include on advocacy pages.

The Advocacy Checklist below provides important questions to consider when analyzing an advocacy page. Application of the general questions from the Checklist of Basic Elements together with the specific questions from the Advocacy Checklist can help a user determine the following:

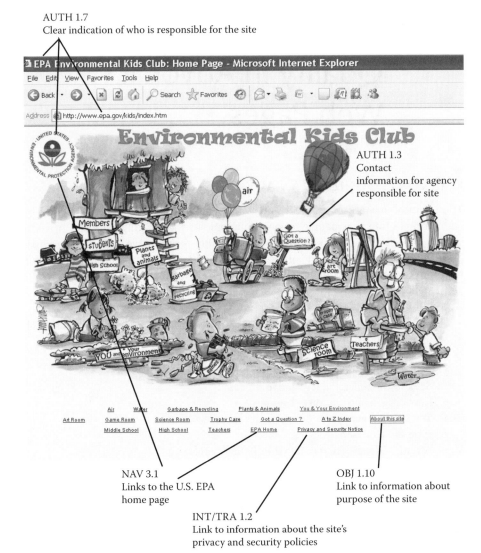

AUTH 1.7
Clear indication of who is responsible for the site

AUTH 1.3
Contact
information for agency
responsible for site

NAV 3.1
Links to the U.S. EPA
home page

OBJ 1.10
Link to information about
purpose of the site

INT/TRA 1.2
Link to information about the site's
privacy and security policies

FIGURE 7.1 An advocacy home page. (Reprinted from U.S. Environmental Protection Agency [EPA], EPA Environmental Kids Club: home page, U.S. EPA, Washington, DC, n.d., http://www.epa.gov/kids/index.htm [accessed April 6, 2009].)

- The nature of the advocacy organization responsible for the contents of the site or pages
- Whether the information on the page is likely to be reliable, authoritative, and trustworthy
- Whether the information on the page is relevant to the user's information needs

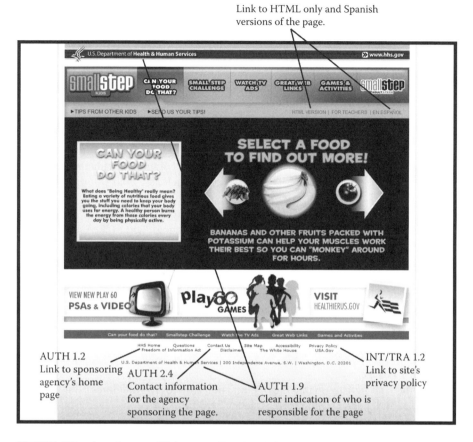

FIGURE 7.2   An advocacy Web page. (Reprinted from U.S. Department of Health and Human Services, Can your food do that? In Smallstep kids, U.S. Department of Health and Human Services, Washington, DC, n.d., http://www.smallstep.gov/kids/flash/can_your_food. html [accessed April 4, 2009].)

These same questions can also be used by a Web author to create advocacy pages that can be recognized as originating from a reliable source.

THE ADVOCACY CHECKLIST: KEYS TO EVALUATING AND CREATING ADVOCACY WEB PAGES

An advocacy Web page is one with the primary purpose of influencing public opinion. The following questions are intended to complement the general questions found on the Checklist of Basic Elements. The greater the number of questions on both the Checklist of Basic Elements and the Advocacy Checklist answered with a "yes," the greater the likelihood that the quality of information on an advocacy Web page can be determined.

If the page you are analyzing is not a home page, it is important on return to the site's home page to answer the questions in the Authority of the Site's Home Page section of the checklist.

AUTHORITY

Authority of the Site's Home Page

The following information should be included either on the site's home page or on a page directly linked to the home page:

- Is there a listing of the names and qualifications of any individuals who are responsible for overseeing the organization (such as a board of directors?) **AUTH 1.7**
- Is there an indication of whether the advocacy organization has a presence beyond the Web? For example, do its members hold face-to-face meetings? **AUTH 1.8**
- Is there an indication whether the site is sponsored by an international, national, or local chapter of an organization? **AUTH 1.9**
- Is there a listing of materials produced by the organization and information about how they can be obtained? **AUTH 1.10**
- Is a complete description of the nature of the organization provided? **AUTH 1.11**
- Is there a statement of how long the organization has been in existence? **AUTH 1.12**
- Is there an indication that the organization adheres to guidelines established by an independent monitoring agency? **AUTH 1.14**
- Is there an indication that the organization has received a tax exemption under section 501(c)(3) of the U.S. Internal Revenue Code? **AUTH 1.19**

ACCURACY

- Are sources for factual information provided, so the facts can be verified in the original source? **ACC1.2**

OBJECTIVITY

- Is there a description of the goals of the person or organization for providing the information? This is often found in a mission statement. **OBJ 1.9**
- Is it clear what issues are being promoted? **OBJ 1.10**
- Are the organization's or person's views on the issues clearly stated? **OBJ 1.11**
- Is there a clear distinction between expressions of opinion on a topic and any informational content that is intended to be objective? **OBJ 1.13**

INTERACTION AND TRANSACTION FEATURES

- For sites with a membership option, is there a mechanism provided for users to become a member of the organization? **INT/TRA 1.4**

8 Keys to Information Quality in Business Web Pages

KEYS TO RECOGNIZING A BUSINESS WEB PAGE

The primary purpose of a business Web page is to promote or sell products or services. Examples of uses for business Web pages include a store selling its products through an online catalog or a computer company providing upgrades for its software and other customer support services via the Web. The URL address of the page often ends in .com (commercial).

A "yes" answer to any of the following questions provides a good indication that the primary purpose of the page is business or marketing. Does the page:

- Promote a product or service?
- Provide customer support?
- Make the company's catalog available online?
- Provide product updates or new versions of a product?
- Provide documentation about a product?
- Request information about a person's lifestyle, demographics, or finances?

ANALYSIS OF BUSINESS WEB PAGES

Figure 8.1 illustrates the home page from the Roots Canada & International Web site. The page demonstrates how many important elements have been included on a business home page. For example, the page

- Displays selected products offered for sale by the company
- Includes a link to an About Roots page that provides a variety of information about the company
- Provides a "contact us" link

Figures 8.2 and 8.3 are two additional pages from the Roots Canada & International Web site that illustrate numerous other important features.

When analyzing a business Web page, it is important first to use the list of general questions found in the Checklist of Basic Elements and subsequently to apply the questions from the Business Checklist. Answering these questions can help a user determine the following:

AUTH 1.1
Company responsible for site
contents clearly indicated

COV/IA 1.7
Listing of products and
services offered

NAV 6.1
Internal search
engine provided

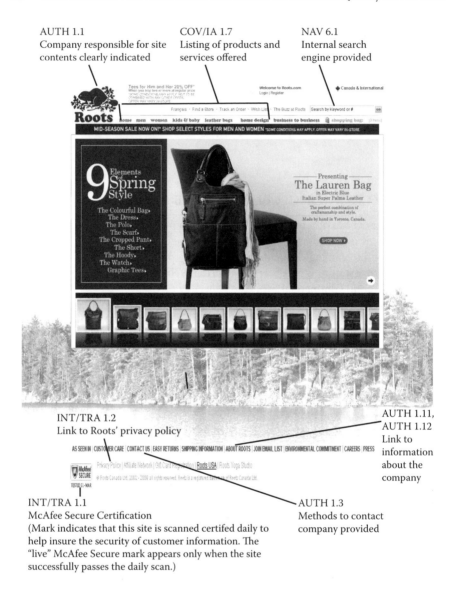

INT/TRA 1.2
Link to Roots' privacy policy

AUTH 1.11,
AUTH 1.12
Link to
information
about the
company

INT/TRA 1.1
McAfee Secure Certification
(Mark indicates that this site is scanned certifed daily to
help insure the security of customer information. The
"live" McAfee Secure mark appears only when the site
successfully passes the daily scan.)

AUTH 1.3
Methods to contact
company provided

FIGURE 8.1   A business home page. (Reprinted from Roots Canada Ltd., Roots Canada & International [home page], 2002–2009-c, http://canada.roots.com/ [accessed March 31, 2009]. Reproduced with permission from Roots Canada Ltd.)

- The nature of the business
- Whether the information at the site is likely to be reliable, authoritative, and trustworthy
- Whether the information at the site is relevant to a user's information needs

These same questions can be used by Web authors as a guide to creating business pages that can be recognized as originating from a reliable, trustworthy source.

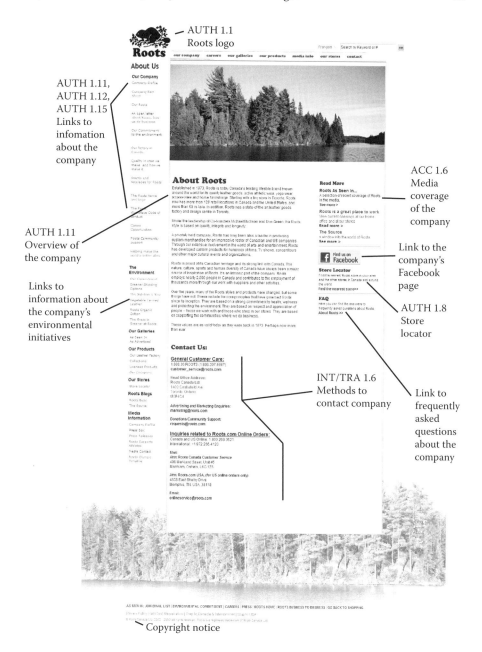

FIGURE 8.2 A business Web page. (Reprinted from Roots Canada Ltd., About us, 2002–2009-a, http://about.roots.com/on/demandware.store/Sites-RootsCorporate-Site/default/Page-Show?cid=ABOUT_US [accessed March 31, 2009]. Reproduced with permission from Roots Canada Ltd.)

INT/TRA 1.2, INT/TRA 1.3
Types of information
collected from visitors to
Roots.com Web site

INT/TRA 1.2
Information about the
company's mailing list

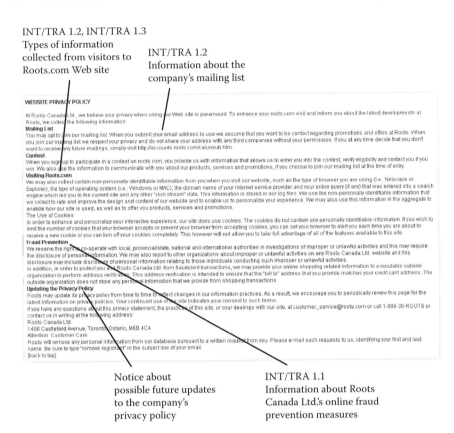

Notice about
possible future updates
to the company's
privacy policy

INT/TRA 1.1
Information about Roots
Canada Ltd.'s online fraud
prevention measures

FIGURE 8.3 Explanation of a business Web site's privacy policy. (Reprinted from Roots Canada Ltd., Privacy policy, 2002–2009-b, http://canada.roots.com/Privacy-Policy-for-Roots/privacyPolicy,default,pg.html [accessed March 31, 2009]. Reproduced with permission from Roots Canada Ltd.)

THE BUSINESS WEB PAGE CHECKLIST: KEYS TO EVALUATING AND CREATING BUSINESS WEB PAGES

The primary purpose of a business Web page is to promote or sell products. The following questions are intended to complement the general questions found on the Checklist of Basic Elements. The greater the number of "yes" answers to questions on both the Checklist of Basic Elements and the Business Checklist, the greater the likelihood that the quality of information on a business Web page can be determined.

If the page you are analyzing is not a home page, it is important to return to the site's home page to answer the questions in the Authority of the Site's Home Page section of the checklist.

AUTHORITY

Authority of the Site's Home Page

The following information should be included either on the site's home page or on a page directly linked to the home page:

- Is it indicated whether the business has a presence beyond the Web? For example, does the business offer a printed catalog or sell its merchandise in a traditional store? **AUTH 1.8**
- Is there a listing of materials about the business, its products, and how the products can be obtained? **AUTH 1.10**
- Is a complete description of the nature of the business and the types of products or services provided? **AUTH 1.11**
- Is there a statement of how long the business has been in existence? **AUTH 1.12**
- Is there a listing of significant employees and their qualifications? **AUTH 1.13**
- Is there an indication that the company adheres to guidelines established by an independent monitoring agency such as the Better Business Bureau? **AUTH 1.14**
- Is financial information about the business provided? **AUTH 1.15**
- For financial information from a public company, is there an indication of whether it has filed periodic reports with the Securities and Exchange Commission (SEC), and is a link provided to the reports? **AUTH 1.16**
- Is any warranty or guarantee information provided for the products or services of the business? **AUTH 1.17**
- Is there a refund policy indicated for any goods purchased from the site? **AUTH 1.18**

ACCURACY

- Is there a link to outside sources, such as product reviews or other independent evaluation of products or servies offered by the business. **ACC 1.6**

OBJECTIVITY

- If there is informational content not related to the company's products or services on the page, is it clear why the company is providing the information? **OBJ 1.8**
- If there is both information-oriented and entertainment-oriented content on the page, is there a clear differentiation between the two? **OBJ 1.14**
- If there is both advertising and entertainment-oriented content on the page, is there a clear differentiation between the two? **OBJ 1.15**

CURRENCY

- If the page includes time-sensitive information, is the frequency of updates described? **CUR 1.5**

COVERAGE AND INTENDED AUDIENCE

- Is there an adequately detailed description for the products and services offered? **COV/IA 1.7**

INTERACTION AND TRANSACTION FEATURES

- Is there a mechanism for users to request additional information from the business, and if so, is there an indication of when they will receive a response? **INT/TRA 1.6**
- Are there clear directions for placing an order for items available from the site? **INT/TRA 1.7**
- Is it clearly indicated when fees are required to access a portion of the site? **INT/TRA 1.8**
- Is it clearly indicated how credit and debit card information will be handled? **INT/TRA 1.10**

9 Keys to Information Quality in Informational Web Pages

KEYS TO RECOGNIZING AN INFORMATIONAL WEB PAGE

The primary purpose of an informational Web page is to provide factual information. Examples of materials found on informational pages include government research reports, census data, and information typically found in encyclopedias and other reference works. Information about a topic can be found on numerous different types of Web pages, so the URL address of an informational page may have any one of a variety of endings.

A "yes" answer to any of the following questions provides a good indication that the primary purpose of the page is informational. Does the page provide the following:

- Factual information about a topic?
- Statistical information?
- The results of research?
- A schedule or calendar of events?
- Transportation schedules?
- Information such as that contained in a reference book?
- A directory of names or businesses?
- A list of course schedules?

ANALYSIS OF INFORMATIONAL WEB PAGES

Figure 9.1 illustrates the home page from the U.S. Food and Drug Administration (FDA) Web site. By looking at the page, we can identify the following characteristics:

- A combination of text and graphics and the site's URL (.gov) indicate who is responsible for the information provided on the page.
- No advertising is present on the page.
- The page includes links to factual and statistical information, the results of research, and other resources related to foods and drugs.

From these factors we can conclude that the page is an informational page.

AUTH 1.1
Agency responsible for site clearly indicated

NAV 4.1
URL for site

NAV 5.1
Site index

NAV 6.1
Internal search engine

Availability of RSS feeds and podcasts

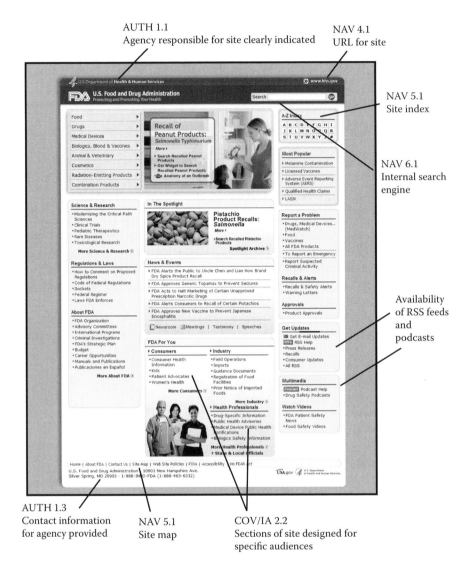

AUTH 1.3
Contact information for agency provided

NAV 5.1
Site map

COV/IA 2.2
Sections of site designed for specific audiences

FIGURE 9.1 An informational home page. RSS, Really Simple Syndication. (Reprinted from U.S. Food and Drug Administration, U.S. Food and Drug Administration [home page], U.S. Food and Drug Administration, Silver Spring, MD, n.d., http://www.fda.gov/default.htm [accessed April 3, 2009].)

When analyzing an informational Web page, the first step is to ask the general questions listed in the Checklist of Basic Elements. In addition, a user must also apply the checklist questions from the Informational Checklist to determine:

- The nature of the information provider.
- Whether the information is likely to be reliable, authoritative, and trustworthy.
- Whether the information at the site is relevant to the user's information needs.

AUTH 1.1
Organization responsible for
page's content clearly indicated

NONTX 1.4
Links to additional software needed
to access portions of site provided

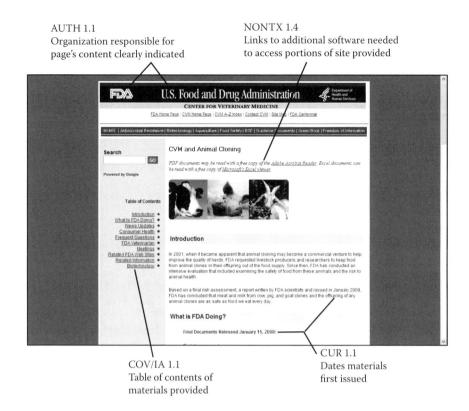

COV/IA 1.1
Table of contents of
materials provided

CUR 1.1
Dates materials
first issued

FIGURE 9.2 An informational Web page. (From U.S. Food and Drug Administration, Center for Veterinary Medicine [CVM], CVM and animal cloning, U.S. Food and Drug Administration, Center for Veterinary Medicine, Rockville, MD, page updated January 31, 2008, http://www.fda.gov/cvm/cloning.htm [accessed April 3, 2009].)

Figure 9.2, also from the U.S. Food and Drug Administration Web site, shows some additional elements that should be incorporated into a well-designed informational page. Meanwhile, Figure 9.3, a page from the U.S. Department of Transportation Research, and Innovative Technology Administration (RITA), Bureau of Transportation Statistics, Web site illustrates features important to include when presenting statistics on an informational page.

THE INFORMATIONAL WEB PAGE CHECKLIST: KEYS TO EVALUATING AND CREATING INFORMATIONAL WEB PAGES

The primary purpose of an informational Web page is to provide factual information. The following questions are intended to complement the general questions found in the Checklist of Basic Elements. The greater the number of "yes" answers to questions on both the Checklist of Basic Elements and the Informational Checklist, the greater the likelihood that the quality of information in an informational Web page can be determined.

Printer friendly
version of table

ACC 1.4
Table clearly
labeled

OBJ 1.4
Page free of advertising

AUTH 1.1
Name of agency responsible for site

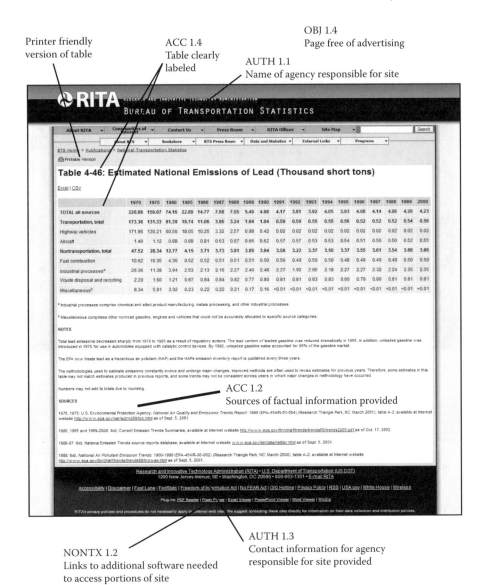

ACC 1.2
Sources of factual information provided

AUTH 1.3
Contact information for agency
responsible for site provided

NONTX 1.2
Links to additional software needed
to access portions of site

FIGURE 9.3 An informational Web page presenting statistics. (Reprinted from U.S. Department of Transportation, Research and Innovative Technology Administration [RITA], Bureau of Transportation Statistics, Table 4-46: Estimated national emissions of lead (thousand short tons), U.S. Department of Transportation, Bureau of Transportation Statistics, Washington, DC, n.d., http://www.bts.gov/publications/national_transportation_statistics/html/table_04_46.html [accessed April 3, 2009].)

If the page you are analyzing is not a home page, it is important to return to the site's home page to answer the questions in the Authority of the Site's Home Page section of the checklist.

AUTHORITY

Authority of the Site's Home Page

The following information should be included either on the site's home page or on a page directly linked to the home page.

If an organization is responsible for providing the information:

- Is there a listing of the names and qualifications of any individuals who are responsible for overseeing the organization (such as a board of directors)? **AUTH 1.7**
- Is there an indication of whether the organization has a presence beyond the Web? For example, does it provide printed materials? **AUTH 1.8**
- Is there a listing of materials produced by the organization and information about how they can be obtained? **AUTH 1.10**
- Is there a listing of significant employees and their qualifications? **AUTH 1.13**

ACCURACY

- If the work is original research by the author, is this clearly indicated? **ACC 1.3**
- Is there an indication that the information has been reviewed for accuracy by an editor or fact-checker or through a peer review process? **ACC 1.5**

CURRENCY

- If the page includes time-sensitive information, is the frequency of updates described? **CUR 1.5**
- If the page includes statistical data, is the date the statistics were collected clearly indicated? **CUR 1.6**
- If the same information is also published in a print source, such as an online dictionary with a print counterpart, is it clear which print edition was the source of the information (i.e., are the title, author, publisher, and date of the print publication listed)? **CUR 1.7**

COVERAGE AND INTENDED AUDIENCE

- Is there a print equivalent to the Web page? If so, is it clear whether the entire work is available on the Web? **COV/IA 1.3**
- If there is a print equivalent to the Web page, is it clear whether the Web version includes additional information not contained in the print version? **COV/IA 1.4**
- If the material is from a work that is out of copyright, is it clear whether and to what extent the material has been updated? **COV/IA 1.5**

10 Keys to Information Quality in News Web Pages

KEYS TO RECOGNIZING A NEWS WEB PAGE

The primary purpose of a news Web page is to provide current information on local, regional, national, or international events. There are also numerous news sites devoted to one particular topic, such as business news, technology news, legal news, and so forth. The site may or may not have a print or broadcast equivalent. For organizations that have a non-Web counterpart, the Web version may or may not duplicate it.

Examples of some organizations with news Web sites include newspapers with a print counterpart, television and radio stations, and Web-based news organizations without a print counterpart. The URL address of a news page frequently ends in .com (commercial).

A "yes" answer to either of the following questions provides a good indication that the primary purpose of the page you are analyzing is to provide news. Does the page:

- Provide current information on local, regional, national, or international events?
- Provide current information on a specific topic such as business, computers, or entertainment?

ANALYSIS OF NEWS WEB PAGES

The home page of the CDC (Centers for Disease Control and Prevention) Online Newsroom (Figure 10.1) and an additional page from the same site (Figure 10.2) provide examples of many of the elements that are important to include on news Web pages:

- The name of the organization responsible for the contents of the site
- The date and time the page was last updated and reviewed
- Contact information for the newsroom staff
- An overview of the topics covered on the page

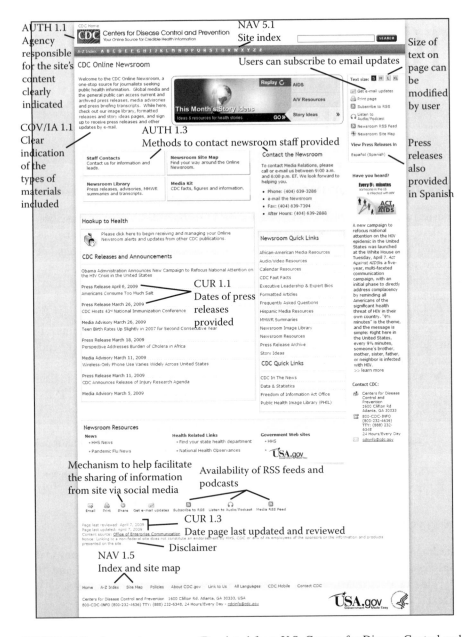

FIGURE 10.1   A news home page. (Reprinted from U.S. Centers for Disease Control and Prevention [CDC], 2009-b, CDC online newsroom [page last updated April 2, 2009], U.S. CDC, Atlanta, GA, http://www.cdc.gov/media/ [accessed April 3, 2009].)

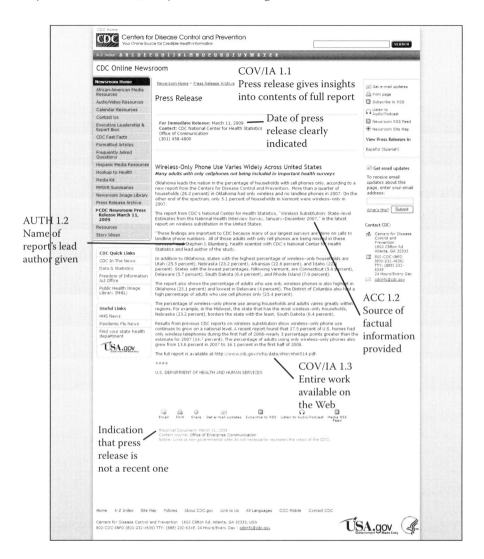

FIGURE 10.2 A news Web page. (Reprinted from U.S. Centers for Disease Control and Prevention [CDC], 2009-a, Wireless-only phone use varies widely across United States, press release, March 11, U.S. CDC, Atlanta, GA, http://www.cdc.gov/media/pressrel/2009/r090311. htm [accessed April 3, 2009].)

- Clear indication of press releases and announcements
- An indication of whether news content provided at the site is available in alternate media formats other than print (e.g., content available via podcasts)

The questions in the News Web Page Checklist complement the general questions listed in the Checklist of Basic Elements. Application of the questions from both checklists to a news Web page can assist a user in determining the following:

- Information about the authority of the news provider
- The extent of news coverage provided at the site and how it differs from any non-Web counterpart
- Whether the news provided at the site is relevant to the user's information needs

THE NEWS WEB PAGE CHECKLIST: KEYS TO EVALUATING AND CREATING NEWS WEB PAGES

The primary purpose of a news Web page is to provide current information on local, regional, national, or international events or to provide news about a particular subject area. The site may or may not have a print or broadcast equivalent. The following questions are intended to complement the general questions found in the Checklist of Basic Elements. The greater the number of "yes" answers to questions on both the Checklist of Basic Elements and the News Web Page Checklist, the greater the likelihood that the quality of information on a news Web page can be determined.

AUTHORITY

Authority of a Page within the Site
- Is there a clear indication if the material has been taken from another source such as a newswire or news service? **AUTH 2.8**

ACCURACY
- Is there an indication that the information has been reviewed for accuracy by an editor or fact-checker? **ACC 1.5**

OBJECTIVITY
- Is there clear labeling of editorial and opinion material? **OBJ 1.12**

CURRENCY
- If the page includes time-sensitive information, is the frequency of updates described? **CUR 1.5**
- If the same information also appears in print, is it clear which print edition the information is from (i.e., national, local, evening, morning edition, etc.)? **CUR 1.7**

- If the material was originally presented in broadcast form, is there a clear indication of the date and time the material was originally broadcast? **CUR 1.8**

Coverage and Intended Audience

- Is there a print equivalent to the Web page or site? If so, is it clear whether the entire work is available on the Web or if parts have been omitted? **COV/IA 1.3**
- If there is a print equivalent to the Web page, is it clear whether the Web version includes additional information not contained in the print version? **COV/IA 1.4**

11 Keys to Information Quality in Personal Web Pages

KEYS TO RECOGNIZING A PERSONAL WEB PAGE

A personal Web page is created by an individual who may or may not be affiliated with a larger institution. Personal pages often are used to showcase an individual's artistic talents, express personal views on a topic, or highlight a favorite hobby or pastime. A personal page can stand alone or be a part of a social networking site like Facebook or MySpace. Actually, blogs can also be considered a unique type of personal page. See Chapter 4 for a more detailed discussion of blogs and other forms of social media.

The URL address of a personal page may have a variety of endings depending on what type of site the page is coming from.

A "yes" answer to any of the following questions provides a good indication that the page you are analyzing is a personal page. Does the page:

- Have as its author a person or family with no official organizational affiliation?
- Consist of a personal expression of something such as:
  - Hobbies or pastimes such as music or sports?
  - Personally authored plays, poems, songs, or other works?
  - Personal opinions on a topic?

ANALYSIS OF A PERSONAL WEB PAGE

Figure 11.1 is an illustration of the home page of a personal Web site, Mave's Media Haven. The inclusion of elements commonly found on other types of Web pages is a normal occurrence on personal Web pages. For example, the Mave's Media Haven home page includes links to the following:

- Advocacy organization pages
- Informational pages
- Entertainment pages
- Business pages

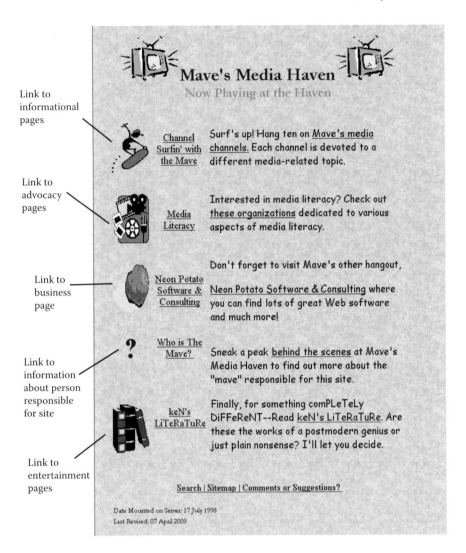

FIGURE 11.1   A personal home page. (Web page created by author.)

To analyze the various pages linked to this home page, it would be necessary to use the Checklist of Basic Elements as well as the appropriate individual checklists.

The "Who is The Mave?" link on the home page leads to information about the creator of Mave's Media Haven site and other background information about the site. This type of information can help the user evaluate the authority and objectivity of the site and its creator.

Use the list of questions found in the Checklist of Basic Elements when analyzing a personal page. Application of the checklist questions to the Mave's Media Haven home page or to any other personal page can help determine the following:

- Who is responsible for the material on the page
- Whether the material on the page is likely to be reliable, authoritative, and trustworthy
- Whether the material at the site is relevant to the user's information needs

12 Keys to Information Quality in Entertainment Web Pages

KEYS TO RECOGNIZING AN ENTERTAINMENT WEB PAGE

The primary purpose of an entertainment Web page is to provide enjoyment to its users by means of humor, games, music, drama, or other similar types of activities. Examples of entertainment Web pages include pages that satirize other Web sites or pages that offer games, jokes, or fan fiction. The URL (uniform resource locator) address of the page may have a variety of endings depending on who is providing the entertainment.

A "yes" answer to any of the following questions provides a good indication that the page is an entertainment page. Does the page:

- Include games or other activities with the primary purpose of providing enjoyment?
- Include music, animation, or video intended primarily to entertain its users?

ENTERTAINMENT PAGES: A NOTE FOR WEB USERS

If the primary purpose of the page is entertainment, enjoy the page. However, pages are not always created merely for entertainment; instead, they may also serve as a vehicle for business, marketing, or educational purposes. Examples of pages that perform these dual roles include ones that:

- Promote a product or service
- Promote a company's public image
- Teach an educational concept
- Promote a TV or radio program or movie

Such additional underlying purposes do not make the entertainment offered less enjoyable. However, Web users should, as they enjoy the entertainment, also be aware of these possible underlying motives and their potential influence on the user.

Knowing why entertainment is provided on a Web site becomes particularly important when children are targeted for marketing a product or for other promotional efforts. In addition, ascertaining the authority of an entertainment provider is important when the payment of fees to the site is involved or when information is collected from a user, either openly via online registration forms and questionnaires,

Players progress is tracked via blinking icons on the map and playing board

Government bodies responsible for game indicated via logos

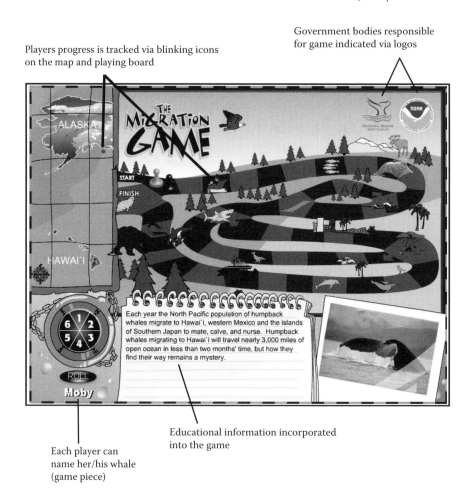

Each player can name her/his whale (game piece)

Educational information incorporated into the game

FIGURE 12.1   Example of blending entertainment and educational content. (Reprinted from U.S. National Oceanic and Atmospheric Administration [NOAA], National Marine Sanctuaries, The migration game, NOAA National Marine Sanctuaries, n.d.-a, http://sanctuaries.noaa.gov/whales/main_page.html [accessed April 7, 2009].)

or transparently through the use of cookies. The procedure for evaluating a page that utilizes entertainment as a tool for promoting something else is similar to that discussed elsewhere in this book:

- Use the Checklist of Basic Elements to evaluate the page.
- Use additional checklists as appropriate.

For example, to evaluate Web pages that combine entertainment and product promotion, after first using the Checklist of Basic Elements, use the Business Checklist to analyze additional business aspects of the pages. When entertainment is used to convey information, consult the Informational Checklist as well as the Checklist of Basic Elements.

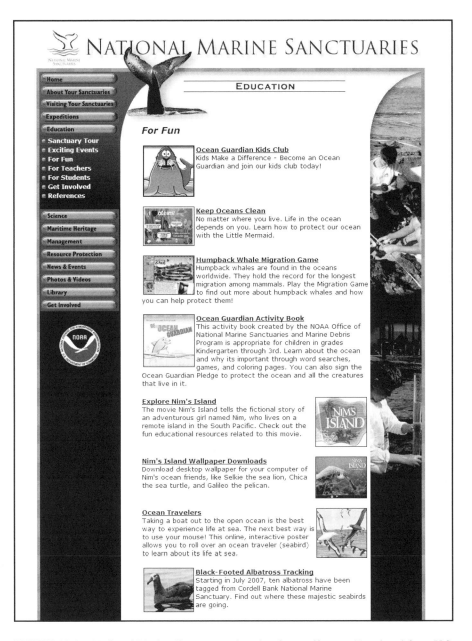

FIGURE 12.2 National Marine Sanctuary education fun stuff page. (Reprinted from U.S. National Oceanic and Atmospheric Administration [NOAA], National Marine Sanctuaries, National Marine Sanctuary education fun stuff, NOAA National Marine Sanctuaries, n.d.-b, revised December 5, 2008, http://sanctuaries.noaa.gov/education/fun/welcome.html [accessed April 7, 2009].)

ANALYSIS OF AN ENTERTAINMENT WEB PAGE

Figure 12.1 shows an entertainment page used for dual purposes—entertainment and education. The entertainment is provided in the form of the Migration Game, a game for one or two players that was inspired by the annual migration of humpback whales in the North Pacific. The first player to successfully finish the migration route wins the game. The game incorporates multimedia elements such as blinking icons that track the progress of each player on both a map and virtual game board displayed side by side on the screen. In addition, players who have sound cards installed on their computers will hear whale and ocean sounds as they play in the game. The Migration Game is just one of a variety of games and other activities available on the National Marine Sanctuaries Education fun stuff page (Figure 12.2) that mix entertainment and education. The page is one component of the National Marine Sanctuaries Web site, a subsite of the U.S. National Oceanic and Atmospheric Administration (NOAA) Web site.

Another popular form of Web entertainment that often serves a dual function is parody. Individuals, businesses, and organizations—indeed, virtually anyone or anything—are apt to be the focus of a parody Web page or site. Just as with other Web-based resources, it is important to determine who is responsible for the parody page or site and its underlying purpose, especially if it also provides informational content or sells products or services.

ENTERTAINMENT WEB PAGE CREATION ISSUES

Creators of entertainment pages that have as their primary purpose promoting enjoyment may not necessarily be concerned with authority, accuracy, currency, objectivity, and coverage issues. However, users should be given, at a minimum, information about who is providing the entertainment and the intended audience. In addition, depending on the type of entertainment offered, it may be necessary for the creator to address other issues as well. The Checklist of Basic Elements can be used as a guide to help in the creation of an entertainment page.

13 Creating Effective Web Pages and Sites

INTRODUCTION

The first part of this chapter offers suggestions to help Web content creators ensure that the material provided at their sites is accessible and easy to use. As stated in Chapter 1, this book does not address visual design issues such as the use of graphics and color. It does, however, address design as it relates to the usability of a page. Any site must be easy enough to use that it does not frustrate its users or otherwise inhibit access to resources offered at the site. It does not matter how much care and attention has gone into creating a site of high information quality if the site is so poorly executed that people are deterred from using it. The chapter also addresses how to effectively facilitate interaction with users of your site and provides checklists relating to the following:

- Consistent and effective use of navigational aids
- Appropriate use of nontext features such as graphics, frames, sound, and video
- Effective handling of interaction and transaction features
- Methods to help ensure your site functions well

The chapter concludes with a brief discussion of metatags and copyright issues.

Following the suggestions outlined in this chapter will help ensure that the Web sites and other Web-based resources you create will not confuse or frustrate users. The actual Web page examples used throughout the book further illustrate how to effectively incorporate these features into Web pages and sites.

THE NAVIGATIONAL AIDS CHECKLIST

Navigational aids are elements that help a user locate information at a Web site and allow the user to easily move from page to page within the site. The greater the number of "yes" answers to the following questions, the more likely the Web page you are creating has effective navigation aids.

NAV 1: Browser Titles

Browser Title for a Home Page
- Does the browser title indicate what company, organization, or person is responsible for the contents of the site? **NAV 1.1**
- Does the browser title indicate that the page is the main, or home page for the site? **NAV 1.2**

- Is the browser title short? **NAV 1.3**
- Is the browser title unique for the site? **NAV 1.4**

Browser Title for Pages That Are Not Home Pages

- Does the browser title indicate what site the page is from? **NAV 1.5**
- Does the browser title clearly describe the contents of the page? **NAV 1.6**
- Is the browser title short? **NAV 1.7**
- Is the browser title unique for the site? **NAV 1.8**
- Does the browser title reflect the location of the page in the site hierarchy? **NAV 1.9**

NAV 2: THE PAGE TITLE

Page Title for a Home Page

- Does the page title describe what site the page is from? This can be done using a logo. **NAV 2.1**
- Does the page title indicate that it is the main, or home page for the site? **NAV 2.2**
- Is the page title short? **NAV 2.3**
- Is the page title unique for the site? **NAV 2.4**

Page Title for a Page That Is Not a Home Page

- Does the page title clearly describe the contents of the page? **NAV 2.5**
- Is the page title short? **NAV 2.6**
- Is the page title unique for the site? **NAV 2.7**
- Does the page title give an indication of the company, organization, or person responsible for the contents of the site? **NAV 2.8**

NAV 3: HYPERTEXT LINKS

- Does the page include a link to the home page? **NAV 3.1**
- Does the page include a link to a site map, index, or table of contents? **NAV 3.2**
- For sites arranged in a hierarchy, does the page include a link to the page one level up in the hierarchy? **NAV 3.3**
- Are internal directional links consistently placed on each page? **NAV 3.4**
- For links that access documents at an external site, is there an indication that the user will be leaving the site? **NAV 3.5**

NAV 4: THE URL FOR THE PAGE

- Does the URL (uniform resource locator) of the page appear in the body of the page? **NAV 4.1**

NAV 6: INTERNAL SEARCH ENGINE

- If your site provides a large amount of information, have you included an internal search engine at the site to enable users to locate specific information quickly and easily? **NAV 6.1**
- Does the internal search engine retrieve complete and appropriate results? **NAV 6.2**

THE NONTEXT FEATURES CHECKLIST

Nontext features include a wide array of elements that require the user to have additional software or a specific browser to utilize the contents of the page. Some examples of nontext features include image maps, sound, video, and graphics. The greater the number of "yes" answers to the following questions, the more likely the Web page you are creating is using nontext features appropriately.

NONTEXT FEATURES (NONTX)

- If the page includes a graphic such as a logo or an image map, is there a text alternative for those viewing the page in text-only mode? **NONTX 1.1**
- If the page includes a nontext file (such as a sound or video file) that may require additional software to play, is there an indication of the additional software needed and where it can be obtained? **NONTX 1.2**
- If a file requires additional software to access it, wherever possible is the same information provided in another format that does not require the additional software? **NONTX 1.3**
- If a page requires a specific browser or a specific version of a browser, does the page specify what is needed and indicate where it can be obtained? **NONTX 1.4**
- When following a link results in the loading of a large graphic, sound, or video file, is information provided to alert the user that this will happen? **NONTX 1.5**
- If animations or other features start automatically when a page is opened, is there a method provided for users to stop them manually? **NONTX 1.6**

THE INTERACTION AND TRANSACTION FEATURES CHECKLIST

Interaction features are mechanisms available at a Web site that enable a user to interact with the person or organization responsible for the site. *Transaction features* are tools that enable a user to enter into a transaction, usually financial, via the site. The greater the number of "yes" answers to the following questions, the more likely it is that your Web site deals appropriately with interaction and transaction features.

INTERACTION AND TRANSACTION ISSUES (INT/TRA)

- If any financial transactions occur at the site, does the site indicate what measures have been taken to ensure their security? **INT/TRA 1.1**
- If the business, organization, or person responsible for the site is requesting information from the user, is there a clear indication of how the information will be used? **INT/TRA 1.2**
- If cookies are used at the site, is the user notified? Is there an indication of what the cookies are used for and how long they last? **INT/TRA 1.3**
- For sites with a membership option, is there a mechanism provided for users to become a member of the organization? **INT/TRA 1.4**
- Is there a feedback mechanism for users to comment about the site? **INT/TRA 1.5**
- Is there a mechanism for users to request additional information from the organization or business, and if so, is there an indication of when they will receive a response? **INT/TRA 1.6**
- Are there clear directions for placing an order for items available from the site? **INT/TRA 1.7**
- Is it clearly indicated when fees are required to access a portion of the site? **INT/TRA 1.8**
- Are any restrictions regarding downloading and other uses of the materials offered on the page clearly stated? **INT/TRA 1.9**

THE WEB SITE FUNCTIONALITY CHECKLIST

Once your Web pages have been created, it is important to check them for accuracy and readability as well as a variety of other factors before you make them public. It is also important to check all links for functionality after the pages are placed on the server and periodically thereafter to make certain that the links continue to function. The following are questions to ask to make sure that your Web site is functioning properly.

PRINTING ISSUES

- Have you checked to make sure pages print out legibly?
- Have any frames been tested to make sure that they can be printed out?
- If a long document has been divided into several different files, have you also made it possible to print out the same document in a single file?

USABILITY AND QUALITY OF EXTERNAL LINKS

- Do you test the functioning of external links when they are first added to your site?
- Do you test the functioning of external links on an ongoing basis to make sure that they continue to link properly?

- Do you check the contents of external links on a regular schedule to make sure that the links are still appropriate for your site and, if currency is an issue, have been kept up to date?

USABILITY OF THE SITE

- Before making your pages public, have you tested them with people who will be using the site and modified the pages accordingly?
- Have you tested the pages to see how they look on as many different browsers as possible? (Whenever possible, create pages so they can be viewed correctly with as many browsers as possible.)
- Do you have a way of soliciting comments from the site's users on a regular basis concerning the layout and content of the site? Do you modify the site accordingly?
- Do you have an ongoing method for testing features at your site to make sure they are all functioning correctly? Features that need regular testing include the following:
  - Internal links
  - External links
  - Forms
  - Images
  - Internal search engines
  - Animation
  - Audio and video clips
- Do you remove outdated material on a regular basis?
- Do you indicate when new additions are placed on your site?
- Do you provide a method for accessing pages that have changed addresses?
- If major revisions have been made to a page, do you indicate what has been revised?
- For any printed documents at your site that have been converted to HTML (Hypertext Markup Language) or PDF (Portable Document Format) files and placed on your site, do you check to make sure that the documents have been converted completely and accurately?
- Do you provide an e-mail address for a "Webmaster" to whom people can write to inform you of any technical problems, such as broken links?

META TAGS

A BRIEF INTRODUCTION

Several HTML tags called *meta tags* may allow Web page authors to exercise some control over the following:

- How the page will be described when it appears in a list of results from a search engine query (*descriptor* meta tags)
- How search engines index the page (*keyword* meta tags)

The meta tags themselves will not be visible to the user viewing the Web page. For example, the information included within a descriptor meta tag is only visible when the Web page appears on a list of results from a search engine or when viewing the HTML source code.

Search engines vary widely in their treatment of meta tags, from using all of the meta information supplied by the page's author to ignoring the meta tags altogether. However, failure to use meta tags would always place pages and sites at the mercy of whatever default formula the search engine uses to index and describe a Web page.

All <META> tags are used within the <HEAD> element of a Web page.

DESCRIPTOR META TAGS

Descriptor meta tags allow Web page authors to provide a description of a Web page or site that can be used by a search engine when it retrieves the page as the result of a query. Failure to use the descriptor meta tag can result in a description of a Web page or site that gives a potential visitor either a poor idea of what the visitor can expect to find at the page or site or, in some cases, no idea at all.

Example of a Descriptor Meta Tag

```
<HEAD>
<TITLE>Using Meta Tags When Creating Web Pages</TITLE>
<META name="description" content="This Web page describes how to use
    meta tags."></HEAD>
```

The following title and description would appear if the page in the example were listed in the results for a search engine query:

Using Meta Tags When Creating Web Pages
This Web page describes how to use meta tags.

KEYWORD META TAGS

A second important use of meta tags involves indexing terms. The Web includes a wide array of search engines, any number of which may index your Web page. However, the methods these search engines use to index pages vary greatly. Some search engines index all the words appearing on a Web page, whereas others index only portions of the page. However, just as the descriptor meta tag allows you to exert some control over how your page is described, the keywords meta tag allows you to supply some keywords that you think best characterize your page. The meta tag keywords will not be visible on your Web page, but they can be used in the indexing process.

Tips for Using the Keyword Meta Tag

- Be sure that the keywords actually describe the materials available on your page.
- Use both common and unique words (i.e., distinctive words that describe your page but few others).
- Use synonyms to supplement words included on your page.
- Provide full names for any important acronyms used on your page.

Example of a Keyword Meta Tag Included with a Descriptor Meta Tag

```
<HEAD>
<TITLE>Using Meta Tags When Creating a Web Page</TITLE>
<META name="description" content="This Web page describes how to use
    the meta tags."
<META name="keywords" content="meta tags, keywords, Web page creation">
</HEAD>
```

This page would be retrieved as the result of a search engine query that used the words *meta tags, keywords,* or *Web page creation* even though only one of these three terms (i.e., *meta tags*) is included in the actual text of the page itself.

COPYRIGHT AND DISCLAIMERS

COPYRIGHT AND THE WEB

The same factors that make the World Wide Web such a convenient channel of information exchange also raise numerous issues about copyright in the Web environment. Many of the questions raised have yet to be answered, and these questions may not be fully resolved by the courts and legislative bodies for years to come. Therefore, with this in mind, only some very general guidelines for Web authors can be offered at present. It should also be noted that the following suggestions pertain to U.S. copyright law; therefore, Web page authors outside the United States should consult the copyright laws for their country. In addition, Web authors in the United States are strongly recommended to consult the U.S. Copyright Office Web site (*www.copyright.gov/*) and other related resources to obtain further information and keep abreast of any future changes in copyright law that are likely to occur. Finally, authors should seek appropriate legal counsel if further advice and clarification on copyright matters is needed.

Barron's *Law Dictionary* defines copyright as "the protection of the works of artists and authors giving them the exclusive right to publish their works or determine who may so publish" (Gifts 1996, 108).

Although copyright protection automatically begins the moment Web content is created, there are several simple steps that authors can take to ensure that they are afforded maximum copyright protection. These steps include the use of copyright notices, copyright registration, and so on.

Works in the Public Domain (Works Not Protected by Copyright)

Copyright protection does not extend to all materials. Large numbers of works lack copyright protection. These materials include the following:

- Works the author has allowed to go into the public domain
- Works for which the copyright has expired
- Works that are authored by the federal government

Although works in these categories may be used without prior permission, it is sometimes hard to determine whether a work falls within the public domain. When uncertainties arise, U.S. Copyright Office records can be searched to ascertain the current copyright status for a particular work.

Fair Use

The term *fair use* refers to a person's right "to use limited portions of" a copyright-protected "work for purposes such as commentary, criticism, news reporting, and scholarly reports" (U.S. Copyright Office 2006). According to the Copyright Act of 1976 (U.S.C. Sect. 107), the following factors should be used to determine whether the use made of a work in any particular case falls under the fair use clause:

1. The purpose and character of the use, including whether such use if of a commercial nature or is for nonprofit educational purposes
2. The nature of the copyrighted work
3. The amount and substantiality of the portion used in relation to the copyrighted work as a whole
4. The effect of the use on the potential market for or value of the copyrighted work

Fair use is yet another hotly debated issue in relation to the Web. Consult resources devoted to copyright issues for more information about the fair use concept as well as to learn of any possible changes to the fair use guidelines.

Copyright Notice

Although use of the copyright notice is not required to obtain copyright protection, it is still a good idea to place it on all of your Web pages. The notice serves as a visible sign to users of your materials that you have claimed ownership of the materials and the rights accompanying the ownership.

Copyright Notice Format

Use the following format when creating a copyright notice:

© or copyright publication date, copyright owner's name

For example:

© 2009 Marsha Ann Tate

Copyright Versus ©

Use the copyright symbol © whenever possible because in some countries the symbol, rather than the word *copyright,* represents the only legally acknowledged form of copyright. This is an especially important concern with the Web because Web materials have the potential for a worldwide audience.

Publication Date

The publication date is the year in which the materials were first created.

Copyright Owner's Name

Although there are various exceptions that allow individuals to use an alias in the copyright notice (if the person is identifiable by that alias), using your full name is probably the least problematic way to identify yourself.

Copyright Registration

Just as copyright notice is not a requirement for copyright protection, neither is registering your copyright with the U.S. Copyright Office. However, registration gives you a far greater opportunity to successfully defend your copyright ownership in any future legal disputes, as well as possibly recoup a larger portion of expenses you may incur in such litigation. If you feel your material is important, take the time to register your copyright with the Copyright Office. Registration information and forms are available at the U.S. Copyright Office's Web site (http://www.copyright.gov/).

Suggested Copyright Guidelines for Web Authors

- Place your copyright notice on every Web page you create.
- Clearly state any additional restrictions you place on the use of your materials (e.g., forbid usage of the materials without your "express permission," etc.).
- Make your copyright notice readable but nonobtrusive.
- Respect the copyright on any works you may include on your Web pages.

Search for your Web pages periodically on various search engines to monitor whether someone may be using your materials without your permission. This can be done by combining a search for the general topic of your page with several distinctive words or phrases that appear on your page. If someone has borrowed your page without permission, the borrowed page may appear in the search results.

A Note on Disclaimers

If a site provides medical or any other type of information that may have potential liability issues, it would be wise to seek legal consultation to determine what type of disclaimer is appropriate for the site.

CREATIVE COMMONS

A growing number of Web authors are using intellectual property-related legal tools provided by Creative Commons (CC) (http://creativecommons.org/), a nonprofit corporation founded in 2001 to provide a "standardized way to grant copyright permissions to their creative works" (Creative Commons n.d.-a;b). Creative Commons provides free copyright licenses together with a Web-based application to creators of works; these creators in turn establish what, if any, restrictions they wish to place on their creations. The CC licenses are not intended to be a substitute for copyright notices and provisions. Instead, the licenses are meant to work in tandem with traditional copyright laws.

Appendix A: Checklist Compilation

THE CHECKLIST OF BASIC ELEMENTS: KEYS TO EVALUATING OR CREATING WEB PAGES

The following questions are general ones that need to be asked when evaluating or creating any Web page, no matter what its type. Answering these questions will help the user determine whether the information on a Web page comes from an authoritative, accurate, and reliable source. The greater the number of "yes" answers, the greater the likelihood that the quality of the information on the page can be determined. The questions can also be used by Web authors as a guide to creating pages that can be recognized as originating from a reliable, trustworthy source.

AUTHORITY (AUTH)

Authority of a Site

The following information should be included either on a site's home page or on a page directly linked to it:

- Is it clear what organization, company, or person is responsible for the contents of the site? This can be indicated by the use of a logo. **AUTH 1.1**
- If the site is a subsite of a larger organization, does the site provide the logo or name of the larger organization? **AUTH 1.2**
- Is there a way to contact the organization, company, or person responsible for the contents of the site? These contact points can be used to verify the legitimacy of the site. Although a phone number, mailing address, and e-mail address are all possible contact points, a mailing address and phone number provide a more reliable way of verifying legitimacy. **AUTH 1.3**
- Are the qualifications of the organization, company, or person responsible for the contents of the site indicated? **AUTH 1.4**
- If all the materials on the site are protected by a single copyright holder, is the name of the copyright holder given? **AUTH 1.5**
- Does the site list any recommendations or ratings from outside sources? **AUTH 1.6**

Authority of a Page

- Is it clear what organization, company, or person is responsible for the contents of the page? Similarity in page layout and design features can help signify responsibility. **AUTH 2.1**

If the material on the page is written by an individual author:

- Is the author's name clearly indicated? **AUTH 2.2**
- Are the author's qualifications for providing the information stated? **AUTH 2.3**
- Is there a way of contacting the author? That is, does the person list a phone number, mailing address, and e-mail address? **AUTH 2.4**
- Is there a way of verifying the author's qualifications? That is, is there an indication of his or her expertise in the subject area or a listing of memberships in professional organizations related to the topic? **AUTH 2.5**
- If the material on the page is copyright protected, is the name of the copyright holder given? **AUTH 2.6**
- Does the page have the official approval of the person, organization, or company responsible for the site? **AUTH 2.7**

Accuracy (ACC)

- Is the information free of grammatical, spelling, and typographical errors? **ACC 1.1**
- Are sources for factual information provided so that the facts can be verified in the original source? **ACC 1.2**
- If there are any graphs, charts, or tables, are they clearly labeled and easy to read? **ACC 1.4**

Objectivity (OBJ)

- Is the point of view of the individual or organization responsible for providing the information evident? **OBJ 1.1**

If there is an individual author of the material on the page:

- Is the point of view of the author evident? **OBJ 1.2**
- Is it clear what relationship exists between the author and the person, company, or organization responsible for the site? **OBJ 1.3**
- Is the page free of advertising? **OBJ 1.4**

For pages that include advertising:

- Is it clear what relationship exists between the business, organization, or person responsible for the contents of the page and any advertisers represented on the page? **OBJ 1.5**

- If there is both advertising and information on the page, is there a clear differentiation between the two? **OBJ 1.6**
- Is there an explanation of the site's policy relating to advertising and sponsorship? **OBJ 1.7**

For pages that have a nonprofit or corporate sponsor:

- Are the names of any nonprofit or corporate sponsors clearly listed? **OBJ 1.16**
- Are links included to the sites of any nonprofit or corporate sponsors so that a user may find out more information about them? **OBJ 1.17**
- Is additional information provided about the nature of the sponsorship, such as what type it is (nonrestrictive, educational, etc.)? **OBJ 1.18**

CURRENCY (CUR)

- Is the date the material was first created in any format included on the page? **CUR 1.1**
- Is the date the material was first placed on the server included on the page? **CUR 1.2**
- If the contents of the page have been revised, is the date (and time, if appropriate) the material was last revised included on the page? **CUR 1.3**
- To avoid confusion, are all dates in an internationally recognized format? Examples of dates in international format (day month year) are 5 June 2009 and 30 April 2010. **CUR 1.4**

COVERAGE AND INTENDED AUDIENCE (COV/IA)

- Is it clear what materials are included on the site? **COV/IA 1.1**
- If the page is still under construction, is the expected date of completion indicated? **COV/IA 1.2**
- If a page incorporates elements of more than one type of page, is there a clear differentiation between the types of content? **COV/IA 1.6**
- Is the intended audience for the material clear? **COV/IA 2.1**
- If material is presented for several different audiences, is the intended audience for each type of material clear? **COV/IA 2.2**

INTERACTION AND TRANSACTION FEATURES (INT/TRA)

- If any financial transactions occur at the site, does the site indicate what measures have been taken to ensure their security? **INT/TRA 1.1**
- If the business, organization, or person responsible for the page is requesting information from the user, is there a clear indication of how the information will be used? **INT/TRA 1.2**
- If cookies are used at the site, is the user notified? Is there an indication of what the cookies are used for and how long they last? **INT/TRA 1.3**
- Is there a feedback mechanism for users to comment about the site? **INT/TRA 1.5**

- Are any restrictions regarding downloading and other uses of the materials offered on the page clearly stated? **INT/TRA 1.9**

THE ADVOCACY CHECKLIST: KEYS TO EVALUATING AND CREATING ADVOCACY WEB PAGES

An advocacy Web page is one with the primary purpose of influencing public opinion. The following questions are intended to complement the general questions found on the Checklist of Basic Elements. The greater the number of questions on both the Checklist of Basic Elements and the Advocacy Checklist answered "yes", the greater the likelihood that the quality of information of an advocacy Web page can be determined.

If the page you are analyzing is not a home page, it is important to return to the site's home page to answer the questions in the Authority of the Site's Home Page section of the checklist.

AUTHORITY

Authority of the Site's Home Page

The following information should be included either on the site's home page or on a page directly linked to the home page:

- Is there a listing of the names and qualifications of any individuals who are responsible for overseeing the organization (such as a board of directors?) **AUTH 1.7**
- Is there an indication of whether the advocacy organization has a presence beyond the Web? For example, do its members hold face-to-face meetings? **AUTH 1.8**
- Is there an indication whether the site is sponsored by an international, national, or local chapter of an organization? **AUTH 1.9**
- Is there a listing of materials produced by the organization and information about how they can be obtained? **AUTH 1.10**
- Is a complete description of the nature of the organization provided? **AUTH 1.11**
- Is there a statement of how long the organization has been in existence? **AUTH 1.12**
- Is there an indication that the organization adheres to guidelines established by an independent monitoring agency? **AUTH 1.14**
- Is there an indication that the organization has received a tax exemption under section 501(c)(3) of the U.S. Internal Revenue Code? **AUTH 1.19**

ACCURACY

- Are sources for factual information provided, so the facts can be verified in the original source? **ACC 1.2**

OBJECTIVITY

- Is there a description of the goals of the person or organization for providing the information? This is often found in a mission statement. **OBJ 1.9**
- Is it clear what issues are being promoted? **OBJ 1.10**
- Are the organization's or person's views on the issues clearly stated? **OBJ 1.11**
- Is there a clear distinction between expressions of opinion on a topic and any informational content that is intended to be objective? **OBJ 1.13**

INTERACTION AND TRANSACTION FEATURES

- For sites with a membership option, is there a mechanism provided for users to become a member of the organization? **INT/TRA 1.4**

THE BUSINESS WEB PAGE CHECKLIST: KEYS TO EVALUATING AND CREATING BUSINESS WEB PAGES

The primary purpose of a business Web page is to promote or sell products. The following questions are intended to complement the general questions found on the Checklist of Basic Elements. The greater the number of "yes" answers to questions on both the Checklist of Basic Elements and the Business Checklist, the greater the likelihood that the quality of information on a business Web page can be determined.

If the page you are analyzing is not a home page, it is important to return to the site's home page to answer the questions in the Authority of the Site's Home Page section of the checklist.

AUTHORITY

Authority of the Site's Home Page

The following information should be included either on the site's home page or on a page directly linked to the home page:

- Is it indicated whether the business has a presence beyond the Web? For example, does the business offer a printed catalog or sell its merchandise in a traditional store? **AUTH 1.8**
- Is there a listing of materials about the business, its products, and how the products can be obtained? **AUTH 1.10**
- Is a complete description of the nature of the business and the types of products or services provided? **AUTH 1.11**
- Is there a statement of how long the business has been in existence? **AUTH 1.12**
- Is there a listing of significant employees and their qualifications? **AUTH 1.13**
- Is there an indication that the business adheres to guidelines established by an independent monitoring agency such as the Better Business Bureau? **AUTH 1.14**
- Is financial information about the business provided? **AUTH 1.15**

- For financial information from a public company, is there an indication of whether it has filed periodic reports with the Securities and Exchange Commission (SEC), and is a link provided to the report? **AUTH 1.16**
- Is any warranty or guarantee information provided for the products or services of the business? **AUTH 1.17**
- Is there a refund policy indicated for any goods purchased from the site? **AUTH 1.18**

ACCURACY

- Is there a link to outside sources such as product reviews or other independent evaluations of products or services offered by the business? **ACC 1.6**

OBJECTIVITY

- If there is informational content not related to the company's products or services on the page, is it clear why the company is providing the information? **OBJ 1.8**
- If there is both information-oriented and entertainment-oriented content on the page, is there a clear differentiation between the two? **OBJ 1.14**
- If there is both advertising and entertainment-oriented content on the page, is there a clear differentiation between the two? **OBJ 1.15**

CURRENCY

- If the page includes time-sensitive information, is the frequency of updates described? **CUR 1.5**

COVERAGE AND INTENDED AUDIENCE

- Is there an adequately detailed description for the products and services offered? **COV/IA 1.7**

INTERACTION AND TRANSACTION FEATURES

- Is there a feedback mechanism for users to comment about the the site? **INT/TRA 1.5**
- Is there a mechanism for users to request additional information from the organization or business, and if so, is there an indication of when they will receive a response? **INT/TRA 1.6**
- Are there clear directions for placing an order for items available from the site? **INT/TRA 1.7**
- Is it clearly indicated when fees are required to access a portion of the site? **INT/TRA 1.8**
- Is it clearly indicated how credit and debit card information will be handled? **INT/TRA 1.10**

THE INFORMATIONAL WEB PAGE CHECKLIST: KEYS TO EVALUATING AND CREATING INFORMATIONAL WEB PAGES

The primary purpose of an informational Web page is to provide factual information. The following questions are intended to complement the general questions found in the Checklist of Basic Elements. The greater the number of "yes" answers to questions on both the Checklist of Basic Elements and the Informational Checklist, the greater the likelihood that the quality of information in an informational Web page can be determined.

If the page you are analyzing is not a home page, it is important to return to the site's home page to answer the questions in the Authority of the Site's Home Page section of the checklist.

AUTHORITY

Authority of the Site's Home Page

The following information should be included either on the site's home page or on a page directly linked to the home page.

If an organization is responsible for providing the information:

- Is there a listing of the names and qualifications of any individuals who are responsible for overseeing the organization (such as a board of directors)? **AUTH 1.7**
- Is there an indication of whether the organization has a presence beyond the Web? For example, does it produce printed materials? **AUTH 1.8**
- Is there a listing of materials produced by the organization and information about how they can be obtained? **AUTH 1.10**
- Is there a listing of significant employees and their qualifications? **AUTH 1.13**

ACCURACY

- If the work is original research by the author, is this clearly indicated? **ACC 1.3**
- Is there an indication that the information has been reviewed for accuracy by an editor or fact-checker or through a peer review process? **ACC 1.5**

CURRENCY

- If the page includes time-sensitive information, is the frequency of updates described? **CUR 1.5**
- If the page includes statistical data, is the date the statistics were collected clearly indicated? **CUR 1.6**
- If the same information is also published in a print source, such as an online dictionary with a print counterpart, is it clear which print edition the information is taken from (i.e., are the title, author, publisher, and date of the print publication listed)? **CUR 1.7**

- Is there a print equivalent to the Web page? If so, is it clear whether the entire work is available on the Web? **COV/IA 1.3**
- If there is a print equivalent to the Web page, is it clear whether the Web version includes additional information not contained in the print version? **COV/IA 1.4**
- If the material is from a work that is out of copyright, is it clear whether and to what extent the material has been updated? **COV/IA 1.5**

THE NEWS WEB PAGE CHECKLIST: KEYS TO EVALUATING AND CREATING NEWS WEB PAGES

The primary purpose of a news Web page is to provide current information on local, regional, national, or international events or to provide news about a particular subject area. The site may or may not have a print or broadcast equivalent.

The following questions are intended to complement the general questions found in the Checklist of Basic Elements. The greater the number of "yes" answers to questions on both the Checklist of Basic Elements and the News Web Page Checklist, the greater the likelihood that the quality of information on a news Web page can be determined.

AUTHORITY

Authority of a Page within the Site
- Is there a clear indication if the material has been taken from another source such as a newswire or news service? **AUTH 2.8**

ACCURACY

- Is there an indication that the information has been reviewed for accuracy by an editor or fact-checker? **ACC 1.5**

OBJECTIVITY

- Is there a clear labeling of editorial and opinion material? **OBJ 1.12**

CURRENCY

- If the page includes time-sensitive information, is the frequency of updates described? **CUR 1.5**
- If the same information also appears in print, is it clear which print edition the information is from (i.e., national, local, evening, morning edition, etc.)? **CUR 1.7**

- If the material was originally presented in broadcast form, is there a clear indication of the date and time the material was originally broadcast? **CUR 1.8**

Coverage and Intended Audience

- Is there a print equivalent to the Web page or site? If so, is it clear whether the entire work is available on the Web or if parts have been omitted? **COV/IA 1.3**
- If there is a print equivalent to the Web page, is it clear whether the Web version includes additional information not contained in the print version? **COV/IA 1.4**

THE NAVIGATIONAL AIDS CHECKLIST

Navigational aids are elements that help a user locate information at a Web site and easily move from page to page within the site. The greater the number of "yes" answers to the following questions, the more likely the Web page you are creating has effective navigational aids.

NAV 1: Browser Titles

Browser Title for a Home Page

- Does the browser title indicate what company, organization, or person is responsible for the contents of the site? **NAV 1.1**
- Does the browser title indicate that the page is the main, or home page for the site? **NAV 1.2**
- Is the browser title short? **NAV 1.3**
- Is the browser title unique for the site? **NAV 1.4**

Browser Title for Pages That Are Not Home Pages

- Does the browser title indicate the source site of the page? **NAV 1.5**
- Does the browser title clearly describe the contents of the page? **NAV 1.6**
- Is the browser title short? **NAV 1.7**
- Is the browser title unique for the site? **NAV 1.8**
- Does the browser title reflect the location of the page in the site hierarchy? **NAV 1.9**

NAV 2: The Page Title

Page Title for a Home Page

- Does the page title describe what site the page is from? This can be done using a logo. **NAV 2.1**

- Does the page title indicate that it is the main, or home page for the site? **NAV 2.2**
- Is the page title short? **NAV 2.3**
- Is the page title unique for the site? **NAV 2.4**

Page Title for a Page That Is Not a Home Page

- Does the page title clearly describe the contents of the page? **NAV 2.5**
- Is the page title short? **NAV 2.6**
- Is the page title unique for the site? **NAV 2.7**
- Does the page title give an indication of the company, organization, or person responsible for the contents of the site? **NAV 2.8**

NAV 3: HYPERTEXT LINKS

- Does the page include a link to the home page? **NAV 3.1**
- Does the page include a link to a site map, index, or table of contents? **NAV 3.2**
- For sites arranged in a hierarchy, does the page include a link to the page one level up in the hierarchy? **NAV 3.3**
- Are internal directional links consistently placed on each page? **NAV 3.4**
- For links that access documents at an external site, is there an indication that the user will be leaving the site? **NAV 3.5**

NAV 4: THE URL FOR THE PAGE

- Does the URL of the page appear in the body of the page? **NAV 4.1**

NAV 5: THE SITE MAP OR INDEX

- Is there a site map or index on the home page or on a page directly linked from the home page? **NAV 5.1**
- Does the site map include at a minimum the main topics at the site? **NAV 5.2**
- Is the site map or index easy to read? **NAV 5.3**
- Is the site map or index organized in a logical manner? **NAV 5.4**
- Are site map and index entries hypertext links to the referenced material? **NAV 5.5**

NAV 6: INTERNAL SEARCH ENGINE

- If your site provides a large amount of information, have you included an internal search engine at the site to enable users to locate specific information quickly and easily? **NAV 6.1**
- Does the internal search engine retrieve complete and appropriate results? **NAV 6.2**

THE NONTEXT FEATURES CHECKLIST

Nontext features include a wide array of elements that require the user to have additional software or a specific browser to utilize the contents of the page. Some examples of nontext features include image maps, sound, video, and graphics. The greater the number of yes answers to the following questions, the more likely the Web page you are creating is using nontext features appropriately.

NONTEXT FEATURES (NONTX)

- If the page includes a graphic such as a logo or an image map, is there a text alternative for those viewing the page in text-only mode? **NONTX 1.1**
- If the page includes a nontext file (such as a sound or video file) that may require additional software to play, is there an indication of the additional software needed and where it can be obtained? **NONTX 1.2**
- If a file requires additional software to access it, wherever possible is the same information provided in another format that does not require the additional software? **NONTX 1.3**
- If a page requires a specific browser or a specific version of a browser, does the page specify what is needed and indicate where it can be obtained? **NONTX 1.4**
- When following a link results in the loading of a large graphic, sound, or video file, is information provided to alert the user that this will happen? **NONTX 1.5**
- If animations or other features start automatically when a page is opened, is there a method provided for users to stop them? **NONTX 1.6**

THE INTERACTION AND TRANSACTION FEATURES CHECKLIST

Interaction features are feedback mechanisms available at a Web site that enable a user to interact with the person or organization responsible for the site. *Transaction features* are tools that enable a user to enter into a transaction, usually financial, via the site. The greater the number of "yes" answers to the following questions, the more likely it is that your Web site deals appropriately with interaction and transaction features.

INTERACTION AND TRANSACTION ISSUES (INT/TRA)

- If any financial transactions occur at the site, does the site indicate what measures have been taken to ensure their security? **INT/TRA 1.1**
- If the business, organization, or person responsible for the site is requesting information from the user, is there a clear indication of how the information will be used? **INT/TRA 1.2**
- If cookies are used at the site, is the user notified? Is there an indication of what the cookies are used for and how long they last? **INT/TRA 1.3**

- For sites with a membership option, is there a mechanism provided for users to become a member of the organization? **INT/TRA 1.4**
- Is there a feedback mechanism for users to comment about the site? **INT/TRA 1.5**
- Is there a mechanism for users to request additional information from the organization or business, and if so, is there an indication of when they will receive a response? **INT/TRA 1.6**
- Are there clear directions for placing an order for items available from the site? **INT/TRA 1.7**
- Is it clearly indicated when fees are required to access a portion of the site? **INT/TRA 1.8**
- Are any restrictions regarding downloading and other uses of the materials offered on the page clearly stated? **INT/TRA 1.9**

THE WEB SITE FUNCTIONALITY CHECKLIST

Once your Web pages have been created, it is important to check them for accuracy and readability as well as a variety of other factors before you make them public. It is also important to check all links for functionality after the pages are placed on the server and periodically thereafter to make certain that the links continue to function. The greater the number of "yes" answers to the following questions, the more likely your Web site is functioning properly.

PRINTING ISSUES

- Have you checked to make sure pages print out legibly?
- Have any frames been tested to make sure that they can be printed out?
- If a long document has been divided into several different files, have you also made it possible to print out the same document in a single file?

USABILITY AND QUALITY OF EXTERNAL LINKS

- Do you test the functioning of external links when they are first added to your site?
- Do you test the functioning of external links on an ongoing basis to make sure that they continue to link properly?
- Do you check the contents of external links on a regular schedule to make sure that the links are still appropriate for your site and, if currency is an issue, have been kept up to date?

USABILITY OF THE SITE

- Before making your pages public, have you tested them with people who will be using the site and modified the pages accordingly?

- Have you tested the pages to see how they look on as many different browsers as possible? (Whenever possible, create pages so they can be viewed correctly with as many browsers as possible.)
- Do you have a way of soliciting comments from the site's users on a regular basis concerning the layout and content of the site? Do you modify the site accordingly?
- Do you have an ongoing method for testing features at your site to make sure they are all functioning correctly? Features that need regular testing include
  - Internal links
  - External links
  - Forms
  - Images
  - Internal search engines
  - Animation
  - Audio and video clips
- Do you remove outdated material on a regular basis?
- Do you indicate when new additions are placed on your site?
- Do you provide a method for accessing pages that have changed addresses?
- If major revisions have been made to a page, do you indicate what has been revised?
- For any printed documents that have been converted to HTML (Hypertext Markup Language) or PDF (Portable Document Format) files and placed on your site, do you check to make sure that the documents have been converted completely and accurately?
- Do you provide an e-mail address for a "Webmaster" to whom people can write to inform you of any technical problems, such as broken links?

Appendix B: Information Quality Questions Compilation

This appendix contains the following:

- Definitions of the eight major categories of information quality elements
- A complete listing by category of questions to consider when evaluating or creating Web pages
- Unique identifiers for each question

AUTHORITY (AUTH)

Definition: The extent to which material is the creation of a person or organization recognized as having definitive knowledge of a given subject area.

QUESTIONS TO ASK ABOUT A SITE'S HOME PAGE

- **AUTH 1.1** Is it clear what organization, company, or person is responsible for the contents of the site? This can be indicated by the use of a logo.
- **AUTH 1.2** If the site is a subsite of a larger organization, does the site provide the logo or name of the larger organization?
- **AUTH 1.3** Is there a way to contact the organization, company, or person responsible for the contents of the site? These contact points can be used to verify the legitimacy of the site. Although a phone number, mailing address, and e-mail address are all possible contact points, a mailing address and phone number provide a more reliable way of verifying legitimacy.
- **AUTH 1.4** Are the qualifications of the organization, company, or person responsible for the contents of the site indicated?
- **AUTH 1.5** If all the materials on the site are protected by a single copyright holder, is the name of the copyright holder given?
- **AUTH 1.6** Does the site list any recommendations or ratings from outside sources?
- **AUTH 1.7** Is there a listing of the names and qualifications of any individuals who are responsible for overseeing the organization (such as a board of directors)?
- **AUTH 1.8** Is there an indication of whether organization or business has a presence beyond the Web? For example, does it hold face-to-face meetings produce printed materials, or have a traditional store?
- **AUTH 1.9** Is there an indication whether the site is sponsored by an international, national, or local chapter of an organization?

- **AUTH 1.10** Is there a listing of materials produced by the organization or business and information about how they can be obtained?
- **AUTH 1.11** Is a complete description of the nature of the organization or business provided?
- **AUTH 1.12** Is there a statement of how long the organization or business has been in existence?
- **AUTH 1.13** Is there a listing of significant employees and their qualifications?
- **AUTH 1.14** Is there an indication that the organization or business adheres to guidelines established by an independent monitoring agency?
- **AUTH 1.15** Is financial information about the business provided?
- **AUTH 1.16** For financial information from a public company, is there an indication of whether it has filed periodic reports with the Securities and Exchange Commission (SEC), and is a link provided to the reports?
- **AUTH 1.17** Is any warranty or guarantee information provided for the products or services offered?
- **AUTH 1.18** Is there a refund policy indicated for any goods purchased from the site?
- **AUTH 1.19** For nonprofit organizations there an indication that the organization has received a tax exemption under section 501(c)(3) of the U.S. Internal Revenue Code?

QUESTIONS TO ASK ABOUT A PAGE THAT IS NOT A HOME PAGE

AUTH 2.1 Is it clear what organization, company, or person is responsible for the contents of the page? Similarity in page layout and design features can help signify responsibility.

If the material on the page is written by an individual author:

- **AUTH 2.2** Is the author's name clearly indicated?
- **AUTH 2.3** Are the author's qualifications for providing the information stated?
- **AUTH 2.4** Is there a way of contacting the author? That is, does the person list a phone number, mailing address, and e-mail address?
- **AUTH 2.5** Is there a way of verifying the author's qualifications? That is, is there an indication of his or her expertise in the subject area or a listing of memberships in professional organizations related to the topic?
- **AUTH 2.6** If the material on the page is copyright protected, is the name of the copyright holder given?
- **AUTH 2.7** Does the page have the official approval of the person, organization, or company responsible for the site?
- **AUTH 2.8** Is there a clear indication if the material has been taken from another source such as a newswire or news service?

ACCURACY (ACC)

Definition: The extent to which information is reliable and free from errors.

QUESTIONS TO ASK

- **ACC 1.1** Is the information free of grammatical, spelling, and typographical errors?
- **ACC 1.2** Are sources for factual information provided, so that the facts can be verified in the original source?
- **ACC 1.3** If the work is original research by the author, is this clearly indicated?
- **ACC 1.4** If there are any graphs, charts, or tables, are they clearly labeled and easy to read?
- **ACC 1.5** Is there an indication that the information has been reviewed for accuracy by an editor or fact-checker or through a peer review process?
- **ACC 1.6** Is there a link to outside sources such as product reviews or other independent evaluations of products or services offered by the business?

OBJECTIVITY (OBJ)

Definition: The extent to which material expresses facts or information without distortion by personal feelings or other biases.

QUESTIONS TO ASK

- **OBJ 1.1** Is the point of view of the individual or organization responsible for providing the information evident?

If there is an individual author of the material on the page:

- **OBJ 1.2** Is the point of view of the author evident?
- **OBJ 1.3** Is it clear what relationship exists between the author and the person, company, or organization responsible for the site?
- **OBJ 1.4** Is the page free of advertising?

For pages that include advertising:

- **OBJ 1.5** Is it clear what relationship exists between the business, organization, or person responsible for the contents of the page and any advertisers represented on the page?
- **OBJ 1.6** If there is both advertising and information on the page, is there a clear differentiation between the two?
- **OBJ 1.7** Is there an explanation of the site's policy relating to advertising and sponsorship?

- **OBJ 1.8** If there is informational content not related to the company's products or services on the page, is it clear why the company is providing the information?
- **OBJ 1.9** Is there a description of the goals of the person or organization for providing the information? This is often found in a mission statement.
- **OBJ 1.10** Is it clear what issues are being promoted?
- **OBJ 1.11** Are the organization's or person's views on the issues clearly stated?
- **OBJ 1.12** Is there clear labeling of editorial and opinion material?
- **OBJ 1.13** Is there a clear distinction between expressions of opinion on a topic and any informational content that is intended to be objective?
- **OBJ 1.14** If there is both information-oriented and entertainment-oriented content on the page, is there a clear differentiation between the two?
- **OBJ 1.15** If there is both advertising and entertainment-oriented content on the page, is there a clear differentiation between the two?

For pages that have a nonprofit or corporate sponsor:

- **OBJ 1.16** Are the names of any nonprofit or corporate sponsors clearly listed?
- **OBJ 1.17** Are links included to the sites of any nonprofit or corporate sponsors so that a user may find out more information about them?
- **OBJ 1.18** Is additional information provided about the nature of the sponsorship, such as what type it is (nonrestrictive, educational, etc.)?

CURRENCY (CUR)

Definition: The extent to which material can be identified as up to date.

QUESTIONS TO ASK

- **CUR 1.1** Is the date the material was first created in any format included on the page?
- **CUR 1.2** Is the date the material was first placed on the server included on the page?
- **CUR 1.3** If the contents of the page have been revised, is the date (and time, if appropriate) the material was last revised included on the page?
- **CUR 1.4** To avoid confusion, are all dates in an internationally recognized format? Examples of dates in international format (day month year) are 5 June 2009 and 30 April 2010.
- **CUR 1.5** If the page includes time-sensitive information, is the frequency of updates described?
- **CUR 1.6** If the page includes statistical data, is the date the statistics were collected clearly indicated or is there a link to the original data?
- **CUR 1.7** If the same information is also published in a print source, such as an online dictionary with a print counterpart, is it clear which print edition

the information is taken from (i.e., are the title, author, publisher, and date of the print publication listed)?

- **CUR 1.8** If the material was originally presented in broadcast form, is there a clear indication of the date and time the material was originally broadcast?

COVERAGE AND INTENDED AUDIENCE (COV/IA)

Definition of coverage: The range of topics included in a work and the depth to which those topics are addressed.

QUESTIONS TO ASK

- **COV/IA 1.1** Is it clear what materials are included on the site?
- **COV/IA 1.2** If the page is still under construction, is the expected date of completion indicated?
- **COV/IA 1.3** Is there a print equivalent to the Web page or site? If so, is it clear whether the entire work is available on the Web or if parts have been omitted?
- **COV/IA 1.4** If there is a print equivalent to the Web page, is it clear whether the Web version includes additional information not contained in the print version?
- **COV/IA 1.5** If the material is from a work that is out of copyright, is it clear whether and to what extent the material has been updated?
- **COV/IA 1.6** If a page incorporates elements of more than one type of page, is there a clear differentiation between the types of content?
- **COV/IA 1.7** Is there an adequately detailed description for the products and services offered?
- **COV/IA 1.8** If the page complements a broadcast or print equivalent to the Web page (i.e., a television show, movie, radio station, etc.) is there an indication of how the broadcast or print equivalent can be accessed?

Definition of intended audience: The group of people for whom the material was created.

QUESTIONS TO ASK

- **COV/IA 2.1** Is the intended audience for the material clear?
- **COV/IA 2.2** If material is presented for several different audiences, is the intended audience for each type of material clear?

INTERACTION AND TRANSACTION FEATURES (INT/TRA)

Definition: Interaction features are mechanisms available at a Web site that enable a user to interact with a person or organization responsible for the site. Transaction features are tools that enable a user to enter into a transaction, usually financial, via the site.

QUESTIONS TO ASK

- **INT/TRA 1.1** If any financial transactions occur at the site, does the site indicate what measures have been taken to ensure their security?
- **INT/TRA 1.2** If the business, organization, or person responsible for the page is requesting information from the user, is there a clear indication of how the information will be used?
- **INT/TRA 1.3** If cookies are used at the site, is the user notified? Is there an indication of what the cookies are used for and how long they last?
- **INT/TRA 1.4** For sites with a membership option, is there a mechanism provided for users to become a member of the organization?
- **INT/TRA 1.5** Is there a feedback mechanism for users to comment about the site?
- **INT/TRA 1.6** Is there a mechanism for users to request additional information from the organization or business, and if so, is there an indication of when they will receive a response?
- **INT/TRA 1.7** Are there clear directions for placing an order for items available from the site?
- **INT/TRA 1.8** Is it clearly indicated when fees are required to access a portion of the site?
- **INT/TRA 1.9** Are any restrictions regarding downloading and other uses of the materials offered on the page clearly stated?
- **INT/TRA 1.10** Is it clearly indicated how credit and debit card information will be handled?

NAVIGATIONAL AIDS (NAV)

Definition: Elements that help a user locate information at a Web site and allow the user to easily move from page to page within the site. Navigational aids may be text, graphics, or a combination of these.

NAV 1: BROWSER TITLES

Questions to Ask for a Home Page
- **NAV 1.1** Does the browser title indicate what company, organization, or person is responsible for the contents of the site?
- **NAV 1.2** Does the browser title indicate that the page is the main, or home page for the site?
- **NAV 1.3** Is the browser title short?
- **NAV 1.4** Is the browser title unique for the site?

Questions to Ask for a Page That Is Not a Home Page
- **NAV 1.5** Does the browser title indicate what site this page is from?
- **NAV 1.6** Does the browser title clearly describe the contents of the page?
- **NAV 1.7** Is the browser title short?

- **NAV 1.8** Is the browser title unique for the site?
- **NAV 1.9** Does the browser title reflect the location of the page in the site's hierarchy?

NAV 2: THE PAGE TITLE

Questions to Ask for a Home Page

- **NAV 2.1** Does the page title describe what site this page is from? This can be done using a logo.
- **NAV 2.2** Does the page title indicate that it is the main, or home page for the site?
- **NAV 2.3** Is the page title short?
- **NAV 2.4** Is the page title unique for the site?

Questions to Ask for a Page That Is Not a Home Page

- **NAV 2.5** Does the page title clearly describe the contents of the page?
- **NAV 2.6** Is the page title short?
- **NAV 2.7** Is the page title unique for the site?
- **NAV 2.8** Does the page title give an indication of the company, organization, or person responsible for the contents of the site?

NAV 3: HYPERTEXT LINKS

Questions to Ask

- **NAV 3.1** Does the page include a link to the home page?
- **NAV 3.2** Does the page include a link to a site map, index, or table of contents?
- **NAV 3.3** For sites arranged in a hierarchy, does the page include a link to the page one level up in the hierarchy?
- **NAV 3.4** Are internal directional links consistently placed on each page?
- **NAV 3.5** For links that access documents at an external site, is there an indication that the user will be leaving the site?

NAV 4: THE URL FOR THE PAGE

Question to Ask

- **NAV 4.1** Does the URL of the page appear in the body of the page?

NAV 5: THE SITE MAP OR INDEX

Questions to Ask

NAV 5.1 Is there a site map or index on the home page or on a page directly linked from the home page?

NAV 5.2 Does the site map or index include at a minimum the main topics at the site?

NAV 5.3 Is the site map or index easy to read?

NAV 5.4 Is the site map or index organized in a logical manner?

NAV 5.5 Are site map and index entries hypertext linked to the referenced material?

NAV 6: INTERNAL SEARCH ENGINE

Questions to Ask

- **NAV 6.1** If the site provides a large amount of information, does it include an internal search engine at the site to enable users to locate specific information quickly and easily?
- **NAV 6.2** Does the internal search engine retrieve complete and appropriate results?

NONTEXT FEATURES (NONTX)

Definition: Nontext features are a wide array of elements that require the user to have additional software or a specific browser to utilize the contents of the page. Some examples include image maps, sound, video, and graphics.

QUESTIONS TO ASK

- **NONTX 1.1** If the page includes a graphic such as a logo or an image map, is there a text alternative for those viewing the page in text-only mode?
- **NONTX 1.2** If the page includes a nontext file (such as a sound or video file) that may require additional software to play, is there an indication of the additional software needed and where it can be obtained?
- **NONTX 1.3** If a file requires additional software to access it, wherever possible is the same information provided in another format that does not require the additional software?
- **NONTX 1.4** If a page requires a specific browser or a specific version of a browser, does the page specify what is needed and indicate where it can be obtained?
- **NONTX 1.5** When following a link results in the loading of a large graphic, sound, or video file, is information provided to alert the user that this will happen?
- **NONTX 1.6** If animations or other features start automatically when a page is opened, is there a method provided for users to stop them manually?

Appendix C: Glossary

Accuracy: The extent to which information is reliable and free from errors.

Advertising: The conveyance of persuasive information about products, services, or ideas using paid announcements, notices, and other methods.

Advertorial: "An advertisement that has the appearance of a news article or editorial in a print publication" (Richards 1995–1996).

Advocacy advertising: Advertising used to promote political or social issues.

Advocacy Web page: A page with the primary purpose of influencing public opinion.

Adware: "A type of software that often comes with free downloads. Some adware displays ads on your computer, while some monitors your computer use (including Web sites visited) and displays targeted ads based on your use" (U.S. Federal Trade Commission et al. n.d.).

Antivirus software (also known as virus protection software): Software designed to protect a computer from computer viruses and, frequently, other types of malware as well (U.S. Federal Trade Commission, et al. n.d.).

Authority: The extent to which material is the creation of a person or organization that is recognized as having definitive knowledge of a given subject area.

Banner advertisement: "A typically rectangular advertisement on a Web site placed above, below, or on the sides of the site's main content and linked to the advertiser's own Web site" (U.S. Department of Education 2003).

Blog.: *See* Weblog.

Bookmark: A URL address stored on a user's computer that allows the user to easily return to a frequently visited Web page. The ability to store bookmarks is a common browser capability.

Browser: Software on a user's computer that permits both the viewing of and navigation among pages on the World Wide Web.

Browser hijacker: "A common spyware program that changes" a "Web browser's home page without the user's knowledge, even if" the user "changes it back" (U.S. Federal Trade Commission, et al. n.d.). *See also* spyware.

Browser title: The title of a Web page that is picked up by the browser from the HTML <TITLE> tag. It usually appears as part of the browser frame at the top of the browser window.

Business Web page: A Web page with the primary purpose of promoting or selling products or services.

Cascading style sheet (CSS): "A mechanism for allowing Web authors and readers to attach" fonts, colors, and other styles "to HTML documents" (U.S. Department of Transportation n.d.).

Channel casting: *See* Web casting

Chat room: A page or section of "a Web site or online service where people can type messages which are displayed almost instantly on the screens of others who are in the 'chat room'" (U.S. Federal Trade Commission, et al. n.d.).

Commercial advertising: "Advertising that involves commercial interests rather than advocating a social or political cause" (Richards 1995–1996). It is designed to sell a specific product or service.

Cookies: Data stored by a Web server on a user's computer. This stored information can be read by the Web server when the user returns to the same site. Cookies enable a business to create a shopping cart into which a person can place items to be purchased, and they also allow a site to tailor Web pages to an individual user's preferences.

Copyright: "A legal term referring to protection granted an individual or organization against the use of an original work without expressed consent" (U.S. National Center for Chronic Disease Prevention and Health Promotion Media Campaign Resource Center 2001).

Corporate sponsor: A business that gives financial or other types of support to something, usually in return for public recognition.

Coverage: The range of topics included in a work and the depth to which those topics are addressed.

CSS: *See* Cascading style sheet

Currency: The extent to which material can be identified as up to date.

Cyber piracy (also known as cyber squatting): Third-party registration of domain names that include the names or trademarks of well-known individuals, businesses, or nonprofit organizations. The third party sometimes alters the domain name "to contain inappropriate and offensive material" or, alternately, withholds the name "for resale" to the individual or organization "for a substantial profit" (U.S. Legal Services Corporation 2007).

Cyber squatting: *See* Cyber piracy

Date of creation: The date material presented on a Web page was first created in any format.

Date last revised: The date material presented on a Web page was last updated.

Date placed on server: The date material presented on a Web page was first placed on the server.

Domain: "A segment of Internet space, denoted by the function or type of information it includes." For example, *.edu* represents space used by an educational institution (U.S. Federal Trade Commission, et al. n.d.).

Encryption: "The scrambling of data into a secret code that can be read only by software set to decode the information" (U.S. Federal Trade Commission, et al. n.d.).

Entertainment Web page: A Web page with the primary purpose of providing enjoyment to its users by means of humor, games, music, drama, or other similar types of activities.

Extensible markup language: *See* XML

Feed reader (also known as an RSS reader): "An application that collects and presents the content provided by a Web feed" (Library of Congress n.d.).

Filter: "Software that screens information on the Internet, classifies its content, and allows the user to block certain kinds of content" (U.S. Federal Trade Commission, et al. n.d.).

Firewall: Hardware or software that helps prevent hackers from gaining access to an individual's computer system (U.S. Federal Trade Commission, et al. n.d.).

Frames: A Web feature that allows the division of a user's browser window into several regions, each of which contains a different Web page. The boundaries between frames may be visible or invisible. Sometimes, each frame can be changed individually, and sometimes one frame in the browser window remains constant while the other frames can be changed by the user.

Freeware: Software that can be downloaded and used free of charge. "The software author retains rights to the program"; thus, it "cannot be resold or re-labeled without the consent of the originator" (U.S. Department of Transportation n.d.). *See also* shareware.

GIF: *See* Graphics Interchange Format

Graphical user interface (GUI): "A computer interface using point-and-click mouse actions (rather than the keyboard exclusively) and pictures (rather than text exclusively)" (U.S. Department of Transportation n.d.).

Graphics: Diagrams, drawings, images, and other types of nontextual material that appear on a Web page.

Graphics Interchange Format (GIF): A popular graphics file format often used on Web pages. "Animated gifs provide a method for adding animation to web pages" (U.S. Department of Transportation n.d.).

GUI: *See* Graphical user interface

Hacker: "Someone who uses the Internet to access computers without permission" (U.S. Federal Trade Commission, et al. n.d.).

Home page: The page at a Web site that serves as the starting point from which other pages at the site can be accessed. A home page serves a function similar to the table of contents of a book.

HTML: *See* Hypertext Markup Language

HTTP: *See* Hypertext Transfer Protocol

Hypertext link (also known as a link): A region of a Web page that, once selected, causes a different Web page or a different part of the same Web page to be displayed. A link can consist of a word or phrase of text, or an image. The inclusion of hypertext links on a Web page allows users to move easily from one Web page to another.

Hypertext Markup Language (HTML): A set of codes that are used to create a Web page. The codes control the structure and appearance of the page when it is viewed by a Web browser. They are also used to create hypertext links to other pages.

Hypertext Transfer Protocol (HTTP): "The standard language that computers connected to the World Wide Web use to communicate with each other" (U.S. Federal Trade Commission, et al. n.d.).

IM: *See* Instant message

Index: A listing, often alphabetical, of the major components of a Web site.

Infomercial: "A commercial that is very similar in appearance to a news program, talk show, or other non-advertising program content. The broadcast equivalent of an advertorial" (Richards 1995–1996).

Informational Web page: A Web page with the primary purpose of providing factual information.

Instant message (IM): "Technology, similar to a chat room, which notifies a user when a friend is online, allowing them to 'converse' by exchanging text messages" (U.S. Federal Trade Commission et al. n.d.).

Institutional advertising: "Advertising used to promote an institution or organization rather than a product or service, in order to create public support and goodwill" (Richards 1995–1996).

Intended audience: The group of people for whom material was created.

Interaction and transaction features: *Interaction features* are mechanisms available at a Web site that enable a user to interact with the person, business, or organization responsible for the site. *Transaction features* are tools that enable a user to enter into a transaction, usually financial, via the site.

Internal search engine: A search engine that searches for words or phrases only within one Web site.

Internet: "A worldwide network of computer networks" (U.S. Federal Financial Institutions Examination Council n.d.).

Internet Protocol (IP): "The computer language that allows computer programs to communicate over the Internet" (U.S. Federal Trade Commission, et al. n.d.).

Internet service provider (ISP): "A company that provides its customers with access to the Internet" (U.S. Federal Financial Institutions Examination Council n.d.).

IP: *See* Internet Protocol

IP address: "An identifier for a computer or device" that "consists of a series of numbers separated by periods" (U.S. Department of Education 2003; U.S. Federal Trade Commission, et al. n.d.).

ISP: *See* Internet service provider

JavaScript: A scripting language (i.e., a relatively simple computer programming language) that can be embedded in the coding of a Web page. JavaScript can be used for animations, sound effects, games, and to cause the text on a page to change when a mouse is moved across it.

Joint Photographic Experts Group: *See* JPEG

JPEG (Joint Photographic Experts Group): A computer file format that reduces the size of or compresses still-image graphics files.

Link: *See* Hypertext link

Malware: A term "used to describe any software designed to cause damage to a single computer, server, or computer network" (U.S. Federal Trade Commission, et al. n.d.).

Meta tags: A group of HTML tags that describe the contents of a Web page. Meta tags do not have to be included on a Web page, and they do not change how the page looks to a user. However, including meta tags on a Web page allows a Web page author to have a certain degree of control over how some search engines index the page.

Microblog: A type of blog wherein the author posts short text entries known as *microposts* to a Web site. Twitter is an example of a microblog.

Multimedia: "Presentation of information that includes video, sound, images, text, animation, and/or other computer generated content" (U.S. Department of Transportation n.d.).

Navigational aids: Elements that help a user locate information at a Web site and allow the user to move easily from page to page within the site. Navigational aids may be text, graphics, or a combination of these.

Netiquette: "The informal rules of Internet courtesy, enforced exclusively by other Internet users" (U.S. Federal Trade Commission, et al. n.d.).

News aggregator: *See* Feed reader

News reader: *See* Feed reader

News Web page: A Web page with the primary purpose of providing current information on local, regional, national, or international events or providing current information about a particular topic, such as business news, legal news, and so forth.

Nonprofit sponsor: An individual or nonprofit organization that provides financial or other types of support for something, usually in return for public recognition.

Nontext features: A wide array of elements that require a user to have additional software or a specific browser to utilize the contents of a Web site. Some examples of nontext features include graphics, image maps, sound, and video.

Objectivity: The extent to which material expresses facts or information without distortion by personal feelings or other biases.

Online profiling: "Compiling information about consumers' preferences and interests by tracking their online movements and actions in order to create targeted ads" (U.S. Federal Trade Commission, et al. n.d.).

Opt in: "When a user explicitly permits a Web site to collect, use, or share his or her information" (U.S. Federal Trade Commission, et al. n.d.).

Opt out: "When a user expressly requests that his/her information not be collected, used, and/or shared. Sometimes a user's failure to 'opt-out' is interpreted as 'opting-in'" (U.S. Federal Trade Commission, et al. n.d.).

P2P: *See* Peer to peer

Page title: The title found in the text of the Web page (as distinguished from the browser title that usually appears at the very top of the screen).

Peer to peer (P2P): "A method of sharing files, usually music, games, or software, with other users through a sharing program that allows uploading and downloading files from other users online" (U.S. Federal Trade Commission, et al. n.d.).

Personal information: Data such as bank and credit card account numbers; Social Security numbers (SSNs); or names, addresses, and phone numbers that can be used to identify specific individuals (U.S. Federal Trade Commission et al. n.d.).

Personal Web page: A page created by an individual who may or may not be affiliated with a larger institution. Personal pages often are used to showcase an individual's artistic talents, express personal views on a topic, or highlight a favorite hobby or pastime. *See also* microblog; weblog.

Phishing: A common online scam by which imposters "send spam or pop-up messages to lure personal information" (e.g., credit card or bank account numbers) "from unsuspecting victims" (U.S. Federal Trade Commission, et al. n.d.).

Plug-ins: Software programs "activated by the Web browser to perform special processing of objects within the HTML document, such as viewing Portable Document Format (PDF) or streaming video objects" (U.S. Department of Health and Human Services n.d.).

Pop-up ads: Unsolicited "advertisements that appear in a separate browser window while a Web site is being viewed" (U.S. Department of Education 2003).

Pop-up messages: *See* Pop-up ads

PDF: *See* Portable Document Format

Portable Document Format (PDF): "A file format developed by Adobe Systems® that captures formatting information from a variety of desktop publishing applications, making it possible to send formatted documents and have them appear on the recipient's monitor or printer as they were intended" (U.S. Department of Education 2003).

Portal: *See* Web portal

Public domain: Items such as books and software that are "available for unrestricted use, and can be copied freely and even renamed and resold" (U.S. Department of Transportation n.d.).

Push Web technology: *See* Web casting

Really Simple Syndication (RSS): Really Simple Syndication's uses include: (a) automatically integrating content from other Web sources into Web sites, (b) updating desired information automatically based on posting date, and (c) "aggregating desired content into one Web site" (U.S. National Archives and Records Administration 2008).

RSS: *See* Really Simple Syndication

RSS reader: *See* Feed reader

Search engine: A tool that can search for words or phrases on a large number of Web pages.

Secure transaction: An encrypted communication between a Web server and a Web browser. Because the data communicated in a secure transaction are encrypted or scrambled, the opportunity for the content to be read by an unauthorized person during the transfer across the Internet is minimized. Financial transactions conducted over the Web are frequently made as secure transactions.

Shareware: "'Try before you buy' software. The author retains full rights to the package. It may be copied at will, but shareware cannot be used at will. There is generally a limited period of use granted without fee, commonly 30 or 60 days. After that period, the user pays a licensing fee to continue using the software" (U.S. Department of Transportation n.d.). *See also* freeware.

Site map: A display, often graphical, of the major components of a Web site.

Social networking sites: "Web sites that allow users to build online profiles; share information, including personal information, photographs, blog entries, and music clips; and connect with other users" (U.S. Federal Trade Commission, et al. n.d.).

Software: "A computer program with instructions that enable the computer hardware to work" (U.S. Federal Trade Commission, et al. n.d.).

Source code: "Instructions to the computer in their original form. Initially, a programmer writes a program in a particular programming language called the source code. To execute the program, the programmer must translate the code into 'machine language,' the only language a computer understands. Source code is the only format readable by humans" (U.S. Department of Education 2003).

Spam: "Unsolicited commercial email, often sent in bulk quantities" (U.S. Federal Trade Commission, et al. n.d.).

Spam zombies: "Computers that have been taken over by spammers without the consent or knowledge of the computer owner. The computers are then used to send spam in a way that hides" its "true origin" (U.S. Federal Trade Commission, et al. n.d.).

Spammer: Someone who sends unsolicited commercial e-mail, often in bulk quantities (U.S. Federal Trade Commission, et al. n.d.).

Sponsorship: Financial or other support given by an individual, business, or organization for something, usually in return for some form of public recognition. *See also* corporate sponsor; nonprofit sponsor.

Spoofing: "A form of masquerading where a trusted IP address is used instead of the true IP address as a means of gaining access to a computer system" (U.S. Federal Financial Institutions Examination Council n.d.).

Spyware: A software program that is secretly installed on a user's computer without his or her consent. Once installed, the program can monitor the user's computer activities, "send pop-up ads, redirect" the computer "to certain Web sites, or record keystrokes, which could lead to identity theft" (U.S. Federal Trade Commission, et al. n.d.).

Surfing: "To move from [Web] site to [Web] site on the Internet in a random or questing way while searching for topics of interest" (U.S. Department of Education 2003).

Target audience: "The specific group that" an "advertiser is attempting to reach and influence" (National Center for Chronic Disease Prevention and Health Promotion Media Campaign Resource Center 2001).

Text attributes: "The color, weight, font, height, and width of text" (U.S. Department of Transportation n.d.).

Text files: "Files that contain no special codes or commands, such as bold, italics, or graphics, only text. Text files … can be read without any special software" (U.S. Department of Transportation n.d.).

Trojans: Software "programs that, when installed on" a "computer, enable unauthorized people to access it and sometimes to send spam from it" (U.S. Federal Trade Commission, et al. n.d.).

TRUSTe: "An online seal program. Web sites displaying the seal have agreed to abide with certain principles regarding user privacy." The site's privacy policy can be viewed by clicking the seal (U.S. Federal Trade Commission, et al. n.d.).

Uniform Resource Locator (URL): A unique identifier that distinguishes a Web page from all other Web pages.

URL: *See* Uniform Resource Locator

Viral advertising (also known as viral marketing): "Marketing techniques that use pre-existing social networks to produce increases in brand awareness, through self-replicating viral processes analogous to the spread of pathological and computer viruses." The techniques "facilitate and encourage people to pass along a marketing message voluntary." There are many different kinds of viral advertising, including text messages, games, images, and audio or video clips (Arizona Office of Tourism n.d.).

Viral marketing: *See* Viral advertising

Virus: "Malicious code that replicates itself within a computer" (U.S. Federal Financial Institutions Examination Council n.d.).

Virus protection software: *See* Antivirus software

Web casting (also known as push Web technology; channel casting): "Technology [that] publishes/broadcasts personalized information to subscribers. Then, instead of using bookmarks and search engines to pull down information, users would run a client application that gets updated with data that is *pushed* down by a server" (U.S. Department of Transportation et al. n.d.).

Web page: An HTML file that has a unique URL address on the World Wide Web.

Web portal: "A Web site or service that offers a broad array of resources and services, such as e-mail, forums, search engines, and online shopping malls." Today, "most of the traditional search engines (e.g., Yahoo®, Google®, etc.) are Web portals, modified to attract and keep a larger audience" (U.S. Department of Education 2003).

Web site: A collection of related Web pages interconnected by hypertext links. Each Web site usually has a home page that provides a table of contents to the rest of the pages at the site.

Web subsite: A site on the World Wide Web that is nested within a larger Web site of a parent organization. The parent organization often has publishing responsibility for the subsite, and the URL for the subsite is usually based on the parent site's URL.

Weblog (also known as a blog): A Web page that functions as a publicly accessible unedited online journal. The journal can be formal or informal in nature and is usually updated on a daily basis by the author (U.S. Department of State n.d.; U.S. Legal Services Corporation 2007).

Wiki: "A Web site that includes the collaboration of work from many different authors. A wiki site often allows anyone to edit, delete, or modify the content of the Web" (U.S. Legal Services Corporation 2007).

Word-of-mouth advertising: The endorsement of a product or service by an individual who has no affiliation with the product or service other than being a user of it and who is not paid for the endorsement.

XML (eXtensible Markup Language): "A metalanguage—a language for describing other languages—which lets you design your own customized markup languages for different types of documents. It is designed to improve the functionality of the Web by providing more flexible and adaptable information identification" (U.S. Federal Financial Institutions Examination Council n.d.).

References

American Society for Magazine Editors (ASME). n.d.-a. About ASME. http://www.magazine. org/Editorial/About_ASME/ (accessed June 21, 2008).

American Society for Magazine Editors (ASME). n.d.-b Best practices for digital media. http://www.magazine.org/ASME/ASME_GUIDELINES/BestPracticesDigMed/index. aspx (accessed April 3, 2009).

Arizona Office of Tourism. n.d. How to develop an interactive marketing strategy. Phoenix, AZ: Arizona Office of Tourism. http://www.azot.gov/documents/Interactive_Marketing. pdf (accessed March 24, 2009).

Cable News Network, Inc. n.d. iReport.com. http://www.ireport.com/index.jspa (accessed April 1, 2009).

Canada. Telecommunications Policy Review Panel. 2006. Telecommunications Policy Review Panel final report, 2006. Ottawa, ON: Telecommunications Policy Review Panel. http:// www.telecomreview.ca/eic/site/tprpgecrt.nsf/vwapj/report_e.pdf/$FILE/report_e.pdf (accessed January 13, 2009).

Consumers Union. 1998–2009. About Consumers Union. Yonkers, NY: Consumers Union. http://www.consumersunion.org/about/ (accessed April 1, 2009).

Copyright Act of 1976. 17 U.S.C. Sect. 101 et seg.

Creative Commons. n.d.-a. History—Creative Commons. San Francisco, CA: Creative Commons. http://creativecommons.org/about/history (accessed March 23, 2009).

Creative Commons. n.d.-b. What is CC—Creative Commons? San Francisco, CA: Creative Commons. http://creativecommons.org/about/what-is-cc (accessed March 23, 2009).

Gifts, S. H. 1996. *Law dictionary* (4th ed.). Hauppauge, NY: Barron's.

Harvey, Ian, Ted Kritsonis, and Grant Buckler. 2007. 39 security threats you need to know about in 2008. *Globe and Mail, TQ Magazine,* November 27, p. 42.

Library of Congress. n.d. Help with Library of Congress RSS feeds. Washington, DC: Library of Congress. http://www.loc.gov/rss/faq.html (accessed January 16, 2008).

Penn State Public Broadcasting. 2004–2009. WPSU/home. University Park, PA: Penn State Public Broadcasting. http://www.wpsu.org/ (accessed March 27, 2009).

Rankin, B. 1998. Best of Tourbus #3: An even closer look at cooks. *The Internet tourbus* [Online], June 20. http://www.tourbus.com/archive/tb063098.htm (accessed August 20, 1998).

Richards, J. 1995–1996. *Dictionary of terminology, advertising.* http://advertising.utexas.edu/ research/terms/index.html (accessed February–March 1998) Web-based pub.

Roots Canada Ltd. 2002–2009a. About us. http://about.roots.com/on/demandware.store/Sites-RootsCorporate-Site/default/Page-Show?cid=ABOUT_US (accessed March 31, 2009).

Roots Canada Ltd. 2002–2009b. Privacy policy. http://canada.roots.com/Privacy-Policy-for-Roots/privacyPolicy,default,pg.html (accessed March 31, 2009).

Roots Canada Ltd. 2002–2009c. Roots Canada & International [home page]. http://canada. roots.com/ (accessed March 31, 2009).

Tate, Marsha Ann, blog. 2008–2009. Web wisdom: How to Create Information Quality on the Web. https://blogs.psu.edu/mt4/mt.cgi (accessed April 2, 2009).

Technorati Inc. 2008. State of the blogosphere, 2008. San Francisco, CA: Technorati. http:// technorati.com/blogging/state-of-the-blogosphere/ (accessed March 30, 2009).

Technorati Inc. n.d. Blogging basics. In *Technorati support: Site guide.* San Francisco, CA: Technorati. http://support.technorati.com/support/siteguide/ (accessed April 2, 2009).

The Math Forum @ Drexel University. 1994–2009-a. About the Math Forum: Mission and history. http://mathforum.org/about.forum.html (accessed April 3, 2009).

The Math Forum @ Drexel University. 1994–2009-b. The Math Forum @ Drexel privacy policy. http://mathforum.org/announce/privacy.html (accessed April 3, 2009).

The Math Forum @ Drexel University. 1994–2009-c. The Math Forum @ Drexel terms of use. http://mathforum.org/announce/terms.html (accessed April 3, 2009).

The Math Forum @ Drexel University. 2008. The Math Forum @ Drexel University: Our thanks. http://mathforum.org/appreciation.html (accessed March 25, 2009).

The Math Forum @ Drexel University. 2009. The Math Forum @ Drexel University [home page]. http://mathforum.org/index.html (accessed April 3, 2009).

The Pennsylvania State University. Department of Plant Pathology. 2009. Department of Plant Pathology [home page] [last modified March 27, 2009]. http://www.ppath.cas.psu.edu/ (accessed April 3, 2009).

United States. The White House. n.d. Subscribe to RSS. Washington, DC: The White House. http://www.whitehouse.gov/rss/ (accessed April 2, 2009).

U.S. Centers for Disease Control and Prevention. 2009-a. Wireless-only phone use varies widely across United States, press release, March 11. Atlanta, GA: U.S. CDC. http://www.cdc.gov/media/pressrel/2009/r090311.htm (accessed April 3, 2009).

U.S. Centers for Disease Control and Prevention. 2009-b. *CDC online newsroom* [page last updated April 2]. Atlanta, GA: U.S. CDC. http://www.cdc.gov/media/ (accessed April 3, 2009).

U.S. Centers for Disease Control and Prevention, the Consumer Product Safety Commission, and the U.S. Fire Administration. n.d. FireSafety.gov for kids [launch page]. http://www.firesafety.gov/kids/flash.shtm (accessed March 26, 2009).

U.S. Copyright Office. Rev. July 12, 2006. Can I use someone else's work? Can someone else use mine? Washington, DC: U.S. Copyright Office. http://www.copyright.gov/help/faq/faq-fairuse.html#locimage (accessed April 5, 2009).

U.S. Department of Agriculture (USDA). Economic Research Service 2008. Food safety and imports: An analysis of FDA food-related import refusal reports [last updated September 9]. [Washington, DC]: USDA Economic Research Service. http://www.ers.usda.gov/Publications/EIB39/ (accessed April 3, 2009).

U.S. Department of Education. Institute of Education Statistics national center for Education Statistics. 2003. Glossary. *In Weaving a Secure Web Around Education: A Guide to Technology Standards and Security*. Washington, DC: U.S. Department of Education, Institute of Education Statistics, National Center for Education Statistics. http://nces.ed.gov/pubs2003/2003381.pdf (accessed June 10, 2007).

U.S. Department of Health and Human Services. n.d. Can your food do that? In *Smallstep kids*. Washington, DC: U.S. Department of Health and Human Services. http://www.smallstep.gov/kids/flash/can_your_food.html (accessed April 4, 2009).

U.S. Department of Health and Human Services. Administration for Children and Families. 2005. Glossary [last updated May 4, 2005]. http://www.acf.hhs.gov/nhsitrc/it_planning/consolidated_definitions/glossary_ref.html (accessed April 2, 2009).

U.S. Department of Transportation. Research and Innovative Technology Administration (RITA). Bureau of Transportation Statistics. n.d. Table 4-46: Estimated national emissions of lead (thousand short tons). Washington, DC: U.S. Department of Transportation, Bureau of Transportation Statistics. http://www.bts.gov/publications/national_transportation_statistics/html/table 04_46.html (accessed April 3, 2009).

U.S. Environmental Protection Agency (EPA). n.d. EPA Environmental Kids Club [home page]. Washington, DC: U.S. EPA. http://www.epa.gov/kids/index.htm (accessed April 6, 2009).

U.S. Federal Financial Institutions Examination Council (FFIEC). n.d. FFIEC Information Technology Examination handbook glossary. Sine loco.: FFIEC. http://www.ffiec.gov/ffiecinfobase/html_pages/gl_01.html (accessed August 20, 2007).

U.S. Federal Housing Administration. 2008-a. FHA wiki [page last changed December 8, 2008]. Washington, DC: U.S. Federal Housing Administration. http://portal.hud.gov/portal/page?_pageid=73,1829262&_dad=portal&_schema=PORTAL (accessed April 2, 2009).

U.S. Federal Housing Administration. 2008-b. Pre qualify. In FHA wiki [page last changed April 29, 2008]. Washington, DC: U.S. Federal Housing Administration. http://portal.hud.gov/portal/page?_pageid=73,1829262&_dad=portal&_schema=PORTAL (accessed April 2, 2009).

U.S. Federal Trade Commission, et al. n.d. Learn the terms. In *OnGuard Online*. Washington, DC: U.S. Federal Trade Commission. http://www.onguardonline.gov/tools/learn-terms.aspx (accessed March 11, 2009).

U.S. Food and Drug Administration. n.d. U.S. Food and Drug Administration [home page]. Silver Spring, MD: U.S. Food and Drug Administration. http://www.fda.gov/default.htm (accessed April 3, 2009).

U.S. Food and Drug Administration, Center for Veterinary Medicine. 2009. CVM and animal cloning [page last updated January 31, 2009]. Rockville, MD: U.S. Food and Drug Administration, Center for Veterinary Medicine (CVM). http://www.fda.gov/cvm/cloning.htm (accessed April 3, 2009).

U.S. Legal Services Corporation, Technology Initiative Grants Program. 2007. Tech glossary. Washington, DC: Legal Services Corporation. http://tig.lsc.gov/techglossary.php (accessed January 19, 2008).

U.S. National Archives and Records Administration. 2008. Implications of recent Web technologies for NARA Web guidance. College Park, MD: U.S. National Archives and Records Administration. http://www.archives.gov (accessed January 19, 2008).

U.S. National Center for Chronic Disease Prevention and Health Promotion media Campaign Resource Center. 2001. Glossary. http://www.cdc.gov/tobacco/mcrc/glossary.htm.

U.S. National Institutes of Health, National Cancer Institute. 2007. Second hand smoke: Questions and answers [reviewed August 1, 2007]. Bethesda, MD: National Cancer Institute. http://www.cancer.gov/cancertopics/factsheet/Tobacco/ETS (accessed April 3, 2009).

U.S. National Oceanic and Atmospheric Administration (NOAA), National Marine Sanctuaries. n.d.-a. The migration game. Silver Spring, MD: NOAA National Marine Sanctuary Program. http://sanctuaries.noaa.gov/whales/main_page.html (accessed April 7, 2009).

U.S. National Oceanic and Atmospheric Administration (NOAA), National Marine Sanctuaries. n.d.-b. National Marine Sanctuary education fun stuff. Silver Spring, MD: NOAA National Marine Sanctuary Program. http://sanctuaries.noaa.gov/education/fun/welcome.html (accessed April 7, 2009).

U.S. National Weather Service, National Hurricane Center. 2007. NHC RSS feeds [page last modified August 16, 2007]. Miami, FL: NOAA/National Weather Service, National Hurricane Center. http://www.nhc.noaa.gov/aboutrss.shtml (accessed January 16, 2008).

U.S. National Weather Service, n.d. NOAA's NWS RSS library [page last modified July 20, 2009]. Silver Spring, MD: National Weather Service. http://www.weather.gov/rss/ (accessed July 27, 2009).

We need you to become an advocate for public television. n.d. http://www.wqln.org/advocate/default.aspx?sid=wpsu (accessed March 27, 2009).

Web site error rocks global oil markets. 2007. Reuters, May 30. http://www.reuters.com/article/internetNews/idUSN3040403520070530 (accessed March 24, 2009).

WebMediaBrands Inc. 2009-a. Dynamic URL. In *Webopedia Computer Dictionary*. Sine loco: WebMediaBrands Inc.

WebMediaBrands, Inc. 2009-b. Static. In *Webopedia Computer Dictionary*. Sine loco: WebMediaBrands Inc.

Worldwide Internet users top 1 billion in 2005. January 4, 2006. Arlington Heights, IL: Computer Industry Almanac. http://www.c-i-a.com/pr0106.htm (accessed March 30, 2009).

Bibliography

3Com Corporation. "Understanding IP Addressing: Everything You Ever Wanted to Know: White Paper." Santa Clara, CA: 3Com Corporation, 2001.

"7 Online Blunders That Threaten Your Identity." *Yahoo! Finance*, August 1, 2008. http://finance.yahoo.com/banking-budgeting/article/105534/7-Online-Blunders-That-Threaten-Your-Identity Personal Finance News from Yahoo! Finance.htm (accessed March 24, 2009).

Adobe. *Adobe® Acrobat® 9 Pro Accessibility Guide: Creating Accessible Forms*. Sine loco: Adobe Systems Inc., n.d. http://www.adobe.com/accessibility/products/acrobat/pdf/A9-creating-accessible-pdf-forms.pdf (accessed March 24, 2009).

Alexander, Janet E., and Marsha Ann Tate. "Teaching Critical Evaluation Skills for World Wide Web Resources." *Computers in Libraries* 16, no. 10 (November–December 1996): 49–55.

American Society for Magazine Editors (ASME). "About ASME." n.d. http://www.magazine.org/Editorial/About_ASME/ (accessed June 21, 2008).

American Society for Magazine Editors (ASME). "Best Practices for Digital Media." n.d. http://www.magazine.org/ASME/ASME_GUIDELINES/BestPracticesDigMed/index.aspx (accessed April 3, 2009).

Anderson, Janna Quitney, and Lee Rainie. "The Future of the Internet III." Sine loco: Pew Internet and American Life Project, December 14, 2008. http://www.pewinternet.org/~/media//Files/Reports/2008/PIP_FutureInternet3.pdf.pdf (accessed March 24, 2009).

Apple. "Apple Glossary." 2008. http://docs.info.apple.com/article.html?artnum=51908 (accessed March 24, 2009).

Arens, William F., Michael F. Weigold, and Christian Arens. *Contemporary Advertising*. 11th ed. Boston, MA: McGraw-Hill Irwin, 2008.

Arizona Office of Tourism. "How to Develop an Interactive Marketing Strategy." Phoenix, AZ: Arizona Office of Tourism, n.d. http://www.azot.gov/documents/Interactive_Marketing.pdf (accessed March 24, 2009).

Armbrust, Michael, Armando Fox, Rean Griffith, et al. "Above the Clouds: A Berkeley View of Cloud Computing." Berkeley, CA: UC Berkeley Reliable Adaptive Distributed Systems Laboratory, February 10, 2009. http://d1smfj0g31qzek.cloudfront.net/abovetheclouds.pdf (accessed March 23, 2009).

Bertolucci, Jeff. "AOL Turns Bebo into One-Stop Inbox." *PC World*, December 10, 2008.

Browne, M. Neil, and Stuart M. Keeley. *Asking the Right Questions: A Guide to Critical Thinking*. 8th ed. Upper Saddle River, NJ: Pearson Prentice Hall, 2007.

Cable News Network Inc. "iReport.com." Sine loco: Cable News Network, n.d. http://www.ireport.com/index.jspa (accessed April 1, 2009).

Canada. Telecommunications Policy Review Panel. "Telecommunications Policy Review Panel final report, 2006." Ottawa, ON: Telecommunications Policy Review Panel, 2006. http://www.telecomreview.ca/eic/site/tprpgecrt.nsf/vwapj/report_e.pdf/$FILE/report_e.pdf (accessed January 13, 2009).

Carvajal, Doreen, and Brad Stone. "New Flavors for Addresses on the Web Are on the Way." *New York Times*, June 27, 2008. http://www.nytimes.com/2008/06/27/technology/27icann.html (accessed March 24, 2009).

Chu, Lenora. "What PayPal Does with Your Money." *CNNMoney.com*, February 26, 2008. http://money.cnn.com/2008/02/26/smbusiness/paypal_float.fsb/index.htm?postversion=2008022611 (accessed March 24, 2009).

ConsumerReports.org. "Online Security Guide." Sine loco: Consumers Union of U.S., Inc., 2004–2009. http://www.consumerreports.org/cro/electronics-computers/resource-center/cyber-insecurity/cyber-insecurity-hub.htm (accessed April 8, 2009).

Consumers Union. "About Consumers Union." Yonkers, NY: Consumers Union, 1998–2009. http://www.consumersunion.org/about/ (accessed April 1, 2009).

Council of Better Business Bureaus. "Start with Trust—Start with BBB" [Better Business Bureau home page]. S.l.: Council of Better Business Bureaus, 2008. http://www.bbb.org/ (accessed April 8, 2009).

Creative Commons. "History—Creative Commons." San Francisco, CA: Creative Commons, n.d. http://creativecommons.org/about/history (accessed March 23, 2009).

Creative Commons. "What Is CC—Creative Commons?" San Francisco, CA: Creative Commons, n.d. http://creativecommons.org/about/what-is-cc (accessed March 23, 2009).

Desautels, Edward. "Software License Agreements: Ignore at Your Own Risk." Washington, DC: US-CERT, 2005, updated 2008. http://www.us-cert.gov/reading_room/EULA.pdf (accessed March 24, 2009).

Digital Media Project. "The Digital Learning Challenge: Obstacles to Educational Uses of Copyrighted Material in the Digital Age: A Foundational White Paper." S.l.: Digital Media Project, August 9, 2006. http://cyber.law.harvard.edu/sites/cyber.law.harvard.edu/files/BerkmanWhitePaper_08-10-2006.pdf (accessed March 24, 2009).

Discovery Education. "Kathy Schrock's Guide for Educators." S.l.: Discovery Education, 2008. http://school.discoveryeducation.com/schrockguide/ (accessed March 24, 2009).

European Publishers Council (EPC). "European Publishers Council: Welcome." 1996–2009. http://www.epceurope.org/index.shtml (accessed March 24, 2009).

Federation of European Direct Marketing. "FEDMA Code on E-commerce & Interactive Marketing." September 2000. http://www.oecd.org/dataoecd/12/21/2091875.pdf (accessed March 24, 2009).

Fishman, Stephen. *The Copyright Handbook: What Every Writer Needs to Know*. 10th ed. Berkeley, CA: Nolo, 2008.

Google. "Blogger: Content Policy." n.d. http://www.blogger.com/content.g (accessed March 23, 2009).

Google. "Digital Millennium Copyright Act–Blogger." n.d. http://www.google.com/blogger_dmca.html (accessed March 23, 2009).

"Google Software Bug Shared Private Online Documents." *Yahoo! News,* March 10, 2009. http://tech.yahoo.com/news/afp/20090310/tc_afp/usitinternetsoftwaregoogle (accessed March 24, 2009).

Harvey, Ian, Ted Kritsonis, and Grant Buckler. "39 Security Threats You Need to Know about in 2008." *Globe and Mail, TQ Magazine,* November 27, 2007, p. 42.

Harvey, Mike. "Twitter to Hit the Big Time with Explosion in Microblogging." *Times Online*, January 22, 2009. http://technology.timesonline,co.uk (accessed January 22, 2009).

Kissel, Richard, ed. "Glossary of Key Information Security Terms." Washington, DC: National Institute of Standards and Technology, April 25, 2006. http://csrc.nist.gov/publications/nistir/NISTIR-7298_Glossary_Key_Infor_Security_Terms.pdf.

Knorr, Eric, and Galen Gruman. "What Cloud Computing Really Means: The Next Big Trend Sounds Nebulous, But It's Not So Fuzzy When You View the Value Proposition from the Perspective of IT Professionals." *InfoWorld*, April 7, 2008. http://www.infoworld.com/article/08/04/07/15FE-cloud-computing-reality_1.html (accessed March 24, 2009).

Lenhart, Amanda. "Adults and Social Network Sites" (Pew Internet Project Data Memo). January 14, 2009. http://www.pewinternet.org/~/media//Files/Reports/2009/PIP_Adult_social_networking_data_memo_FINAL.pdf.pdf (accessed March 24, 2009).

Lenhart, Amanda. "Twitter and Status Updating" (Pew Internet Project Data Memo). February 12, 2009. http://www.pewinternet.org/~/media//Files/Reports/2009/PIP%20Twitter%20Memo%20FINAL.pdf (accessed March 24, 2009).

Lenhart, Amanda, and Mary Madden. "Teens, Privacy & Online Social Networks: How Teens Manage Their Online Identities and Personal Information in the Age of MySpace." Washington, DC: Pew Internet & American Life Project, April 18, 2007. http://www.pewinternet.org/~/media//Files/Reports/2007/PIP_Teens_Privacy_SNS_Report_Final.pdf.pdf (accessed March 24, 2009).

Lenhart, Amanda, Mary Madden, Alexandra Rankin Macgill, and Aaron Smith. "Teens and Social Media: The Use of Social Media Gains a Greater Foothold in Teen Life as They Embrace the Conversational Nature of Interactive Online Media." Washington, DC: Pew Internet & American Life Project, December 19, 2007. http://www.pewinternet.org/~/media//Files/Reports/2007/PIP_Teens_Social_Media_Final.pdf.pdf (accessed March 24, 2009).

Library of Congress. "Help with Library of Congress RSS Feeds." Washington, DC: Library of Congress, n.d. http://www.loc.gov/rss/faq.html (accessed January 16, 2008).

Madden, Mary, Susannah Fox, Aaron Smith, and Jessica Vitak. "Digital Footprints: Online Identity Management and Search in the Age of Transparency." Washington, DC: Pew Internet & American Life Project, December 2007. http://pewinternet.org/~/media//Files/Reports/2007/PIP_Digital_Footprints.pdf.pdf (accessed March 24, 2009).

Malone, Michael S. "Is Google Turning into Big Brother?" *ABC News*, September 5, 2008. http://abcnews.go.com/Business/story?id=5727509&page=1 (accessed March 24, 2009).

Markoff, John. "Microsoft Plans 'Cloud' Operating System." *NYTimes.com*, October 28, 2008. http://www.nytimes.com/2008/10/28/technology/28soft.html (accessed March 24, 2009).

McCarthy, Caroline. "Oops! Twitter Phishing Scam Snares CNN Anchor." *CNET News*, January 5, 2009.

McMillan, Robert. "Researcher Accuses Sears of Spreading Spyware." *PC World*, January 2, 2008.

Michigan State Police (MSP). "MSP—Glossary of Computer & Internet Terms." East Lansing, MI: State of Michigan, 2001–2009. http://www.michigan.gov/msp/0,1607,7-123-1589_1711_4579-142199--,00.html (accessed March 24, 2009).

National Cyber Security Alliance. "Welcome to Stay Safe Online." http://www.staysafeonline.org/ (accessed March 24, 2009).

Pegoraro, Rob. "Facebook Backs into a 'Bill of Rights.'" *Washington Post*, February 19, 2009. http://www.washingtonpost.com/wp-dyn/content/story/2009/02/19/ST2009021901903.html (accessed March 24, 2009).

Pennsylvania State University. Department of Plant Pathology. "Department of Plant Pathology" [home page]. Last modified March 27, 2009. http://www.ppath.cas.psu.edu/ (accessed April 3, 2009).

Richardo, J. "Dictionary of Terminology, Advertising." 1995–1996. http://advertising.vtexas.edu/research/terms/index.html (accessed February–March 1998).

Roots Canada Ltd. "About Us." 2002–2009. http://about.roots.com/on/demandware.store/Sites-RootsCorporate-Site/default/Page-Show?cid=ABOUT_US (accessed March 31, 2009).

Roots Canada Ltd. "Privacy Policy." 2002–2009. http://canada.roots.com/Privacy-Policy-for-Roots/privacyPolicy,default,pg.html (accessed March 31, 2009).

Roots Canada Ltd. "Roots Canada & International Home Page." 2002–2009. http://canada.roots.com/ (accessed March 31, 2009).

Sandoval, Greg. "Who's to Blame for Spreading Phony Jobs Story?" *CNET News*, October 4, 2008. http://news.cnet.com/8301-1023_3-10058419-93.html (accessed October 10, 2008).

Schwartz, Ephraim. "The Dangers of Cloud Computing: On-Demand Apps and Services Have Several Security Risks that IT Should Address Up Front." *InfoWorld* July 7, 2008. http://www.infoworld.com/article/08/07/07/28NF-cloud-computing-security_1.html (accessed March 24, 2009).

Silverblatt, Art. *Media Literacy: Keys to Interpreting Media Messages.* 3rd ed. Westport, CT: Praeger, 2008.

Story, Louise, and Brad Stone. "Facebook Retreats on Online Tracking." *New York Times*, November 30, 2007. http://www.nytimes.com/2007/11/30/technology/30face.html (accessed March 24, 2009).

Technorati Inc. "Blogging Basics." In *Technorati Support: Site Guide.* San Francisco, CA: Technorati, n.d. http://support.technorati.com/support/siteguide/ (accessed April 2, 2009).

Technorati Inc. "State of the Blogosphere, 2008." San Francisco, CA: Technorati, 2008. http://technorati.com/blogging/state-of-the-blogosphere/ (accessed March 30, 2009).

Tellis, Gerard, and Tim Ambler, eds. *The Sage Handbook of Advertising.* Los Angeles, CA: Sage Publications, 2007.

The Math Forum @ Drexel. "About the Math Forum." 1994–2009. http://mathforum.org/about.forum.html (accessed March 24, 2009).

The Math Forum @ Drexel University. "The Math Forum @ Drexel University" [home page]. 2009. http://mathforum.org/index.html (accessed April 3, 2009).

The Math Forum @ Drexel University. "The Math Forum @ Drexel University: Our Thanks." 2008. http://mathforum.org/appreciation.html (accessed March 25, 2009).

The Math Forum @ Drexel University. "The Math Forum @ Drexel Privacy Policy." 1994–2009. http://mathforum.org/announce/privacy.html (accessed April 3, 2009).

The Math Forum @ Drexel University. "The Math Forum @ Drexel Terms of Use." 1994–2009. http://mathforum.org/announce/terms.html (accessed April 3, 2009).

United States Computer Emergency Readiness Team (US-CERT). "Home Computer Security." Pittsburgh, PA: Carnegie Mellon University, 2002. http://www.us-cert.gov/reading_room/HomeComputerSecurity/home_computer_security.pdf (accessed June 24, 2008).

United States Copyright Office. "Circular 1: Copyright Basics." Washington, DC: U.S. Copyright Office, rev. July 2008. http://www.copyright.gov/circs/circ1.pdf (accessed March 23, 2009).

United States Copyright Office. "U.S. Copyright Office" [home page]. Washington, DC: U.S. Copyright Office, rev. March 10, 2009. http://www.copyright.gov/ (accessed March 24, 2009).

U.S. Centers for Disease Control and Prevention. *CDC Online Newsroom.* Atlanta, GA: U.S. CDC, page last updated April 2, 2009. http://www.cdc.gov/media/ (accessed April 3, 2009).

U.S. Centers for Disease Control and Prevention. "Wireless-Only Phone Use Varies Widely Across United States," press release. Atlanta, GA: U.S. CDC, March 11, 2009. http://www.cdc.gov/media/pressrel/2009/r090311.htm (accessed April 3, 2009).

U.S. Department of Education. Institute of Education Sciences. National Center for Education Statistics. "Glossary." In *Weaving a Secure Web Around Education: A Guide to Technology Standards and Security.* Washington, DC: U.S. Department of Education, Institute of Education Sciences, National Center for Education Statistics, 2003. http://nces.ed.gov/pubs2003/2003381.pdf (accessed June 10, 2007).

U.S. Department of Health and Human Services. "Can Your Food Do That?" In *Smallstep Kids.* Washington, DC: U.S. Department of Health and Human Services, n.d. http://www.smallstep.gov/kids/flash/can_your_food.html (accessed April 4, 2009).

U.S. Department of Health and Human Services. "Help with RSS." In *PandemicFlu.gov.* Washington, DC: U.S. Department of Health and Human Services, n.d. http://www.pandemicflu.gov/rss/whatisRSS.html (accessed August 2, 2007).

U.S. Department of Health and Human Services. Administration for Children and Families. "Glossary." Washington, DC: U.S. Department of Health and Human Services, Administration for Children and Families, last updated May 4, 2005. http://www.acf.hhs.gov/ nhsitrc/it_planning/consolidated_definitions/glossary_ref.html (accessed October 5, 2007).

U.S. Department of Transportation. Federal Lands Highway Division. "Glossary." In *Design Visualization Guide*. Washington, DC: U.S. Department of Transportation, Federal Lands Highway Division, n.d. http://www.efl.fhwa.dot.gov/manuals/dv/glossary/ (accessed January 16, 2008).

U.S. Department of Transportation. Research and Innovative Technology Administration (RITA). Bureau of Transportation Statistics. "Table 4-46: Estimated National Emissions of Lead (Thousand Short Tons)." Washington, DC: U.S. Department of Transportation, Bureau of Transportation Statistics, n.d. http://www.bts.gov/publications/national_transportation_statistics/html/table 04_46.html (accessed April 3, 2009).

U.S. Environmental Protection Agency (EPA). "EPA Environmental Kids Club: Home Page." Washington, DC: U.S. EPA, n.d. http://www.epa.gov/kids/index.htm (accessed April 6, 2009).

U.S. Federal Financial Institutions Examination Council (FFIEC). *FFIEC Information Technology Examination Handbook Glossary*. Sine loco: FFIEC, n.d. http://www.ffiec. gov/ffiecinfobase/html_pages/gl_01.html (accessed August 20, 2007).

U.S. Federal Trade Commission, Plaintiff, v. John Zuccarini, Individually and d/b/a Cupcake Party, Cupcake-Party, … Defendant. "Judgment and Permanent Injunction." U.S. District Court for the Eastern District of Pennsylvania, April 9, 2002. http://www.ftc. gov/os/2002/05/johnzuccarinijudandpi.pdf (accessed March 24, 2009).

U.S. Federal Trade Commission et al. "Learn the Terms." In *OnGuard Online*. Washington, DC: United States, Federal Trade Commission et al., n.d. http://www.onguardonline. gov/tools/learn-terms.aspx (accessed April 2, 2009).

U.S. Food and Drug Administration. "U.S. Food and Drug Administration home page." Silver Spring, MD: U.S. Food and Drug Administration, n.d. http://www.fda.gov/default.htm (accessed April 3, 2009).

U.S. Food and Drug Administration. Center for Veterinary Medicine. "CVM and Animal Cloning." Rockville, MD: U.S. Food and Drug Administration, Center for Veterinary Medicine (CVM), page updated January 31, 2008. http://www.fda.gov/cvm/cloning.htm (accessed April 3, 2009).

U.S. National Archives and Records Administration (NARA). "Implications of Recent Web Technologies for NARA Web Guidance." College Park, MD: U.S. National Archives and Records Administration, 2008. http://www.archives.gov (accessed January 19, 2008).

U.S. National Archives and Records Administration. "NARA Guidance on Managing Web Records." College Park, MD: U.S. National Archives and Records Administration, 2005. http://www.archives.gov/records-mgmt/pdf/managing-web-records-index.pdf (accessed January 12, 2009).

U.S. National Center for Chronic Disease Prevention and Health Promotion Media Campaign Resource Center. "Glossary." 2001. http://www.cdc.gov/tobacco/mcrc/glossary.htm.

U.S. National Oceanic and Atmospheric Administration (NOAA). National Marine Sanctuaries. "The Migration Game." Silver Spring, MD: NOAA National Marine Sanctuary Program, n.d. http://sanctuaries.noaa.gov/whales/main_page.html (accessed April 7, 2009).

U.S. National Oceanic and Atmospheric Administration (NOAA). National Marine Sanctuaries. "National Marine Sanctuary Education Fun Stuff." Silver Spring, MD: NOAA National Marine Sanctuary Program, n.d. http://sanctuaries.noaa.gov/education/fun/welcome. html (accessed April 7, 2009).

US-CERT. "Recognizing and Avoiding Email Scams." Washington, DC: US-CERT, 2005, updated October 2008. http://www.us-cert.gov/reading_room/emailscams_0905.pdf (accessed March 24, 2009).

US-CERT. "Spyware." Washington, DC: US-CERT, 2005, updated October 2008. http://www. us-cert.gov/reading_room/spywarehome_0905.pdf (accessed March 24, 2009).

US-CERT. "Virus Basics." Washington, DC: US-CERT, n.d. http://www.uscert.gov/reading_room/virus.html (accessed March 24, 2009).

"Web Site Error Rocks Global Oil Markets." Reuters, May 30, 2007. http://www.reuters.com/article/internetNews/idUSN3040403520070530 (accessed March 24, 2009).

"Your Data Love It or Lose It." *Baseline.com*, January 31, 2008.

Index